高等职业教育新目录新专标电子与信息大类教材

容器云服务架构与运维

刘晓玲　刘洪海　主　编
贾　强　陈　晨　李　超　副主编

电子工业出版社
Publishing House of Electronics Industry
北京·BEIJING

内 容 简 介

本书以 Docker 和 Kubernetes 为重点,从 Docker 的基本使用入手,深入浅出地讲解了容器云基础组件 Docker 的构建、操作、技术原理和实际使用过程中的典型案例,进而详细介绍了 Kubernetes 的基本原理、安装部署、资源管理等内容,最后通过持续集成的典型案例讲解了 Kubernetes 的最佳实践。全书共分为初识 Docker 容器、Docker 镜像仓库、持久化存储卷、容器网络、容器镜像、Docker Compose 编排服务、Docker Swarm 编排服务、Kubernetes 集群服务、基于 K8S 的 CI/CD 服务等 9 个单元。

本书可以作为云计算技术应用专业、计算机网络技术专业及其他计算机相关专业的云计算课程教材,也可以作为云计算相关培训教材,还可以作为 1+X 证书"云计算平台运维与开发"的配套教材,也可以供云计算相关从业人员和广大计算机爱好者自学使用。

未经许可,不得以任何方式复制或抄袭本书之部分或全部内容。
版权所有,侵权必究。

图书在版编目(CIP)数据

容器云服务架构与运维/刘晓玲,刘洪海主编. —北京:电子工业出版社,2023.11
ISBN 978-7-121-45667-1

Ⅰ.①容⋯ Ⅱ.①刘⋯ ②刘⋯ Ⅲ.①云计算—高等学校—教材 Ⅳ.①TP393.027

中国国家版本馆 CIP 数据核字(2023)第 097128 号

责任编辑:康 静
印 刷:山东华立印务有限公司
装 订:山东华立印务有限公司
出版发行:电子工业出版社
　　　　　北京市海淀区万寿路 173 信箱　邮编:100036
开　本:787×1092　1/16　印张:12　字数:300.8 千字
版　次:2023 年 11 月第 1 版
印　次:2024 年 12 月第 2 次印刷
定　价:39.00 元

凡所购买电子工业出版社图书有缺损问题,请向购买书店调换。若书店售缺,请与本社发行部联系,联系及邮购电话:(010)88254888,88258888。
质量投诉请发邮件至 zlts@phei.com.cn,盗版侵权举报请发邮件至 dbqq@phei.com.cn。
本书咨询联系方式:(010)88254609 或 hzh@phei.com.cn。

前　　言

在互联网时代，为了实现应用更快地开发迭代和更好地弹性伸缩，互联网应用不再采用传统的三层架构，而是使用微服务的方式来实现软件系统的松耦合、跨部门开发及快速交付等目的。

与此同时，应用运行的底层系统环境也发生了天翻地覆的变化，从硬件服务型到虚拟机再到私有云，从本地数据中心到公有云。应用架构的颗粒化、底层系统的复杂化给应用的开发、测试、发布和运维都带来了巨大的挑战，应用服务器的数量成倍增长，环境交互速度的要求也越来越高，应用迁移也越来越频繁。传统的基础架构平台已经跟不上互联网的步伐。

在这样的时代背景下，Docker 容器技术和 Kubernetes 容器管理平台应运而生。它打破了传统的部署架构。Docker 定义了容器技术的标准，使得应用的部署落地变得十分简单，应用可以稳定地运行在容器中。同时 Kubernetes 平台能保证容器按照用户的期望状态运行着。随着 Docker 和 Kubernetes 技术的发展，绝大多数有条件的公司都已经在使用 Kubernetes 部署和发布自己的线上业务了。对一名普通开发人员来说，这一切可能发生得太快，让人搞不清楚它是怎么发生的，也会疑惑 Docker 和 Kubernetes 之间到底是个什么关系。

本书由院校教师和南京第五十五所技术开发有限公司、国基北盛（南京）科技发展有限公司、山东浪潮优派科技教育有限公司等企业云计算专家与高级工程师，共同设计与编写，全面地介绍了 Docker 容器技术与 Kubernetes 平台的优点、优势、使用场景与使用方法。我们对分散在不同文献中的理论和概念进行了汇总和整理，并结合大量实操案例，介绍了 Docker 容器服务和 Kubernetes 平台的安装、配置与使用，为读者提供了完整的 Docker 容器服务与 Kubernetes 平台的使用指南。

1. 本书面向的读者

（1）高职院校或职教本科计算机或电子信息大类专业的学生。

（2）云计算、容器领域的初级从业人员。

（3）对 Docker 容器与 Kubernetes 有兴趣并有志从事该领域工作的人。

2. 本书的主要内容

单元 1 对 Docker 容器概念、主要功能、基本概念等做了介绍，并通过实操案例介绍了 Docker 的安装与基本命令使用。

单元 2 对 Docker 镜像仓库概念、作用、使用场景、优点做了详细的介绍，并通过实操案例，介绍了 Registry 和 Harbor 两种镜像仓库的安装与使用。

单元 3 介绍了容器中几种实现持久化存储卷的方法以及不同存储卷的使用场景。

单元 4 对容器中的几种网络模式进行了详细的介绍，并说明了不同网络模式适用于何种场景。

单元 5 介绍了容器镜像的架构以及两种制作容器镜像的方式与优缺点。

单元 6 对容器编排工具 Compose 进行了介绍，通过真实案例，让读者掌握 Compose 编排部署应用的方式与方法。

单元 7 对另一个容器编排工具 Swarm 进行了介绍，通过实战案例让读者掌握 Swarm 的构建与使用方法。

单元 8 介绍了 Kubernetes 容器云管理平台的基本概念、与 Docker 的区别等。通过对 Kubernetes 容器云平台的部署与使用体验，让读者掌握 Kubernetes 平台中常用模块的使用与 Yaml 模板文件的编写。

单元 9 介绍了一个 Kubernetes 的最佳实践案例，让读者了解什么是 CI/CD、CI/CD 的工作流程、应用场景与优势等。

本书由济南职业学院刘晓玲、刘洪海担任主编，贾强、陈晨、李超担任副主编。全书由刘晓玲、刘洪海拟定大纲、统稿和修改。具体分工为：单元 1、单元 6、单元 7、单元 8 由刘晓玲、刘洪海编写；单元 2、单元 3、单元 4、单元 5 由贾强、陈晨、李超编写。济南职业学院赵光勤、贾丽洁参与了部分单元实训任务的编写。

编者

目　　录

单元 1　初识 Docker 容器 ·· 1
　　学习目标 ·· 1
　　1.1　认识 Docker ·· 1
　　　　1.1.1　什么是 Docker ·· 1
　　　　1.1.2　为什么要使用 Docker ·· 4
　　　　1.1.3　Docker 容器的几个概念 ··· 5
　　1.2　Docker 的安装与使用 ·· 8
　　归纳总结 ·· 20
　　课后练习 ·· 20
　　技能训练 ·· 20

单元 2　Docker 镜像仓库 ·· 21
　　学习目标 ·· 21
　　2.1　Docker 仓库的基本概念 ··· 21
　　　　2.1.1　Registry 私有镜像仓库 ··· 21
　　　　2.1.2　Harbor 企业级镜像仓库 ··· 22
　　2.2　Docker 镜像仓库的部署和使用 ·· 23
　　归纳总结 ·· 32
　　课后练习 ·· 33
　　技能训练 ·· 33

单元 3　持久化存储卷 ··· 34
　　学习目标 ·· 34
　　3.1　认识容器存储卷 ·· 34
　　3.2　实现存储卷持久化 ·· 37
　　归纳总结 ·· 48
　　课后练习 ·· 48
　　技能训练 ·· 49

单元 4　容器网络 ··· 50
　　学习目标 ·· 50

- 4.1 容器网络的几种原型 ··· 50
  - 4.1.1 容器网络概述 ··· 50
  - 4.1.2 容器的 4 种网络模式 ··· 51
- 4.2 实现容器的多种网络模式 ··· 53
- 归纳总结 ··· 58
- 课后练习 ··· 58
- 技能训练 ··· 59

单元 5 容器镜像 ··· 60
- 学习目标 ··· 60
- 5.1 制作容器镜像的方式 ··· 60
- 5.2 容器镜像制作 ··· 62
- 归纳总结 ··· 67
- 课后练习 ··· 67
- 技能训练 ··· 68

单元 6 Docker Compose 编排服务 ··· 69
- 学习目标 ··· 69
- 6.1 认识 Docker Compose 编排服务 ··· 69
- 6.2 使用 Docker Compose 编排服务 ··· 70
- 归纳总结 ··· 93
- 课后练习 ··· 93
- 技能训练 ··· 93

单元 7 Docker Swarm 编排服务 ··· 94
- 学习目标 ··· 94
- 7.1 认识 Docker Swarm 编排服务 ··· 94
- 7.2 使用 Docker Swarm 编排服务 ··· 97
- 归纳总结 ··· 115
- 课后练习 ··· 116
- 技能训练 ··· 116

单元 8 Kubernetes 集群服务 ··· 117
- 学习目标 ··· 117
- 8.1 认识 Kubernetes 集群服务 ··· 117
  - 8.1.1 Kubernetes 介绍 ··· 117
  - 8.1.2 Kubernetes 发展 ··· 120
- 8.2 使用 Kubernetes 集群服务 ··· 125
- 归纳总结 ··· 152
- 课后练习 ··· 152

技能训练 ……………………………………………………………………… 153

单元 9　基于 K8S 的 CI/CD 服务 ……………………………………………… 154
　　学习目标 ……………………………………………………………………… 154
　　9.1　CI/CD 服务 ……………………………………………………………… 154
　　　　9.1.1　CI/CD 服务介绍 ………………………………………………… 154
　　　　9.1.2　CI/CD 的工作流程与使用场景 ………………………………… 155
　　　　9.1.3　CI/CD 的优点 …………………………………………………… 156
　　9.2　基于 K8S 构建持续集成 ……………………………………………… 157
　　归纳总结 ……………………………………………………………………… 180
　　课后练习 ……………………………………………………………………… 180
　　技能训练 ……………………………………………………………………… 180

单元 1 初识 Docker 容器

学习目标

通过本单元的学习，要求了解什么是 Docker 容器技术、Docker 技术的起源与发展，了解 Docker 容器技术与虚拟机 VM 的区别和优势，了解 Docker 容器技术的几个重要概念等。本单元培养读者掌握安装 Docker 容器服务与使用基本命令、本地上载 Docker 镜像与拉取公有镜像、使用不同参数和不同容器镜像启动容器等技能；培养读者的动手实操、独立思考和解决问题的能力。

1.1 认识 Docker

1.1.1 什么是 Docker

1. 容器技术介绍

在介绍 Docker 容器技术之前，先讲一讲什么是容器。IT 行业的"容器技术"是英文单词"Linux Container"的直译。Container 这个单词有集装箱、容器的含义（主要偏集装箱意思）。不过，在中文环境下，咱们要交流要传授，如果翻译成"集装箱技术"就有点拗口，所以结合中国人的吐字习惯和文化背景，更喜欢用容器这个词。不过，如果要形象地理解 Linux Container 技术的话，念成集装箱会比较好。海边码头里的集装箱是运载货物用的，它是一种按规格制造标准化的钢制箱子。集装箱的特色，在于其格式一致，并可以层层重叠，所以可以大量放置在特别设计的远洋轮船中（早期航运是没有集装箱概念的，那时候货物杂乱无章地堆放，很影响出货和运输效率）。有了集装箱，就可以更加快捷方便地为生产商提供廉价的运输服务。

因此，IT 世界里借鉴了这一理念。早期，人们都认为硬件抽象层基于 Hypervisor（系统管理程序，一种运行在基础物理服务器和操作系统之间的中间软件层，可允许多个操作系统和应用共享硬件）的虚拟化方式可以最大限度地提供虚拟化管理的灵活性。各种不同操作系统的虚拟机都能通过 Hypervisor（KVM、XEN 等）来衍生、运行、销毁。然而，随着时间推移，用户发现采用 Hypervisor 这种方式的麻烦越来越多。为什么？因为对于 Hypervisor 环境来说，每个虚拟机都需要运行一个完整的操作系统以及其中安装好的大量应用程序。但在实际生产开发环境里，用户更关注的是自己部署的应用程序，如果每次部署发布用户都得搞一个完整的操作系统和附带的依赖环境，那么就会让任务变得很重，性能变得很低下。

基于上述情况，人们就在想，有没有其他方式能让人更加关注应用程序本身，底层的操作系统和环境可以共享和复用？换句话说，那就是用户部署一个服务并成功运行后，若再想将其移植到另外一个地方，可以不用再安装一套操作系统和依赖环境。这就像集装箱运载一样，用户把货物如一辆兰博基尼跑车（好比开发好的应用 App），打包放到一个集装箱里，它通过货轮可以轻而易举地从"上海码头"（CentOS7.2 环境）运送到"纽约码头"（Ubuntu14.04 环境）。而且运输期间，用户的兰博基尼（App）没有受到任何的损坏（文件没有丢失），在另外一个码头卸货后，依然可以正常启动。

2008 年，Linux Container 容器技术的诞生就解决了 IT 世界里"集装箱运输"的问题。Linux Container（简称 LXC）是一种内核轻量级的操作系统层虚拟化技术。Linux Container 主要由 Namespace 和 Cgroup 两大机制来保证实现。那么 Namespace 和 Cgroup 是什么呢？刚才提到了集装箱，集装箱的作用当然是可以对货物进行打包隔离了，不让 A 公司的货跟 B 公司的货混在一起，不然卸货时就分不清楚了。Namespace 也具有一样的作用——隔离。但光有隔离还不够，还需要对货物进行资源的管理。同样，航运码头也有这样的管理机制：货物用什么规格的集装箱，货物用多少个集装箱，将哪些货物优先运走，遇到极端天气怎么暂停运输服务、怎么改航道，等等。与此机制对应的 Cgroup 就负责资源管理控制作用，比如进程组使用 CPU/MEM 的限制，进程组的优先级控制，进程组的挂起和恢复，等等。

2. Docker 技术介绍

当前，Docker 几乎是容器的代名词，很多人以为 Docker 就是容器。其实，这是错误的认识，除了 Docker 还有 CoreOS。所以，容器世界里并不是只有 Docker 一家。既然不是一家就很容易出现分歧。任何技术的出现都需要一个标准来规范它，不然各搞各的很容易导致技术实现的碎片化，出现大量的冲突和冗余。因此，在 2015 年，由 Google、Docker、CoreOS、IBM、微软、红帽等厂商联合发起的 OCI（Open Container Initiative）组织成立了，并于 2016 年 4 月推出了第一个开放容器标准。该标准主要包括 Runtime 运行时标准和 Image 镜像标准。该标准的推出，为成长中的市场带来了稳定性，让企业能放心采用容器技术，用户在打包、部署应用程序后，可以自由选择不同的容器 Runtime；同时，镜像打包、建立、认证、部署、命名也都能按照统一的规范来做。

Docker 是一个开源的应用容器引擎，基于 Go 语言并遵从 Apache2.0 协议开源。Docker 可以让开发者打包用户的应用以及依赖包到一个轻量级、可移植的容器中，然后发布到任何流行的 Linux 机器上，也可以实现虚拟化。容器完全使用沙箱机制，相互之间不会有任何接口，更重要的是容器的性能开销极低。Docker 改变了虚拟化的方式，使开发者可以直接将自己的成果放入 Docker 中进行管理。方便快捷已经是 Docker 的最大优势，过去需要用数天乃至数周的任务，在 Docker 容器的处理下，只需要数秒就能完成。一方面，云计算时代的到来，使开发者不必为了追求效果而配置高额的硬件，Docker 改变了高性能必然高价格的思维定势。另一方面，Docker 与云空间的结合，让云空间得到更充分的利用，不仅解决了硬件管理的问题，也改变了虚拟化的方式。

3. Docker 与虚拟机的对比

在进行 Docker 与虚拟机的对比之前，可以看一下 Docker 官方关于两者的对比，如

图 1-1 所示。

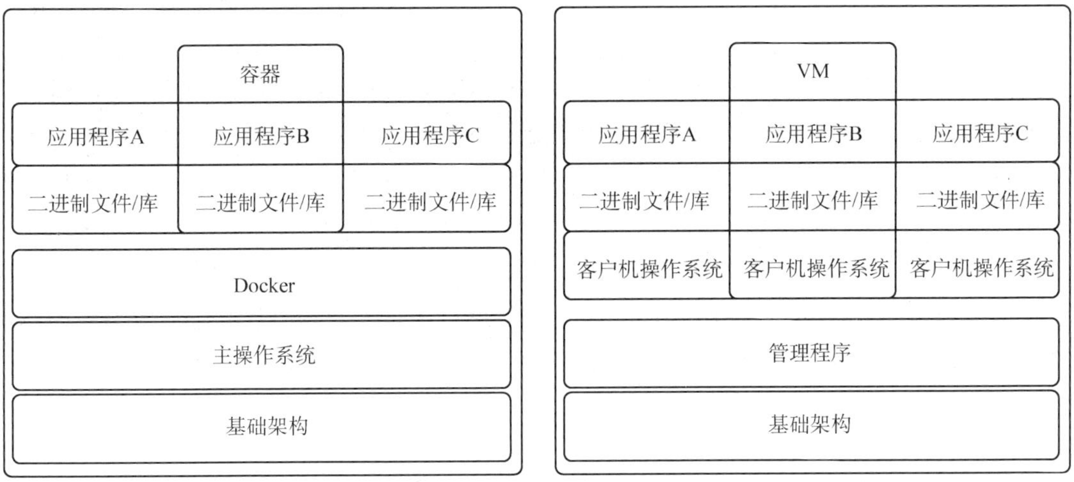

容器			VM		
应用程序A	应用程序B	应用程序C	应用程序A	应用程序B	应用程序C
二进制文件/库	二进制文件/库	二进制文件/库	二进制文件/库	二进制文件/库	二进制文件/库
			客户机操作系统	客户机操作系统	客户机操作系统
Docker			管理程序		
主操作系统					
基础架构			基础架构		

图 1-1 容器与虚拟机对比图

容器是应用程序层的抽象，将代码和依赖项打包在一起。多个容器可以在同一台计算机上运行，并与其他容器共享 OS 内核，每个容器在用户空间中作为隔离的进程运行。容器占用的空间少于 VM（容器镜像的大小通常为几十 MB），可以处理更多的应用程序，并且需要的 VM 和操作系统更少。

虚拟机（VM）是将一台服务器转变为多台服务器的物理硬件的抽象。虚拟机管理程序允许多个 VM 在单台计算机上运行。每个 VM 包含客户操作系统、应用程序、必要的二进制文件和库的完整副本，占用空间大小为数十 GB。VM 也可能启动缓慢。

用一句话来总结：Docker 容器比虚拟机轻量多了！

对比虚拟机与 Docker，Docker 守护进程可以直接与主操作系统进行通信，为各个 Docker 容器分配资源；它还可以将容器与主操作系统隔离，并将各个容器互相隔离。虚拟机启动需要数分钟，而 Docker 容器可以在数毫秒内启动。由于没有臃肿的从操作系统，Docker 可以节省大量的磁盘空间以及其他系统资源。虽然 Docker 有众多优势，但读者也没有必要完全否定虚拟机技术，因为两者有不同的使用场景。虚拟机更擅长于彻底隔离整个运行环境。例如，云服务提供商通常采用虚拟机技术隔离不同的用户。而 Docker 通常被用于隔离不同的应用，例如前端、后端以及数据库。

对比服务器虚拟化与 Docker，服务器好比运输码头，拥有场地和各种设备（服务器硬件资源）；服务器虚拟化好比码头上的仓库，拥有独立的空间以堆放各种货物或集装箱（仓库之间完全独立，独立的应用系统和操作系统）；Docker 好比集装箱，负责各种货物的打包（将各种应用程序和所依赖的运行环境打包成标准的容器，容器之间互相隔离）。Docker 有着小巧、迁移部署快速、运行高效等特点，但隔离性比服务器虚拟化差，正如不同的集装箱属于不同的运单（Docker 上运行不同的应用实例），相互独立（隔离），但由同一个库管人员管理（主操作系统内核），因此通过库管人员可以看到所有集装箱的相关信息（因为共享操作系统内核，因此相关信息会共享）。服务器虚拟化就好比在码头上（物理主机及虚拟化层），建立了多个独立的"小码头"——仓库（虚拟机），其拥有完全独立（隔离）的空间，属于不同的客户（虚拟机所有者）。每个仓库有各自的库管人员（当前虚拟机的操作系统内核），无法管理其他仓库，不存在信息共享的情况。因此，需要根据不同的应用场景和需求，采用不同的方式使用 Docker 技术或使用服务器虚拟化技术。例如，一个典型的 Docker 应用场景是主机上的 Docker 实例属于单一用户，此时使用 Docker 可以在保证安全的同时充分发挥 Docker 的技术优势。对于隔离要求较高的环境如混合用户环境，就可以使用服务器虚拟化技术。

1.1.2 为什么要使用 Docker

作为一种新兴的虚拟化方式，Docker 跟传统的虚拟化方式相比具有众多的优势，具体介绍如下。

1. 更高效地利用系统资源

由于容器不需要进行硬件虚拟以及运行完整的操作系统等额外开销，所以 Docker 对系统资源的利用率更高。无论是应用执行速度、内存损耗，还是文件存储速度，都要比传统虚拟机技术更高效。因此，相比虚拟机技术，一个配置相同的主机，往往可以运行更多数量的 Docker 应用。

2. 更短的启动时间

传统的虚拟机技术启动应用服务往往需要数分钟，而 Docker 容器应用由于直接运行于宿主内核，无须启动完整的操作系统，因此可以做到秒级，甚至毫秒级的启动时间，大大地节约了开发、测试、部署的时间。

3. 一致的运行环境

开发过程中一个常见的问题是环境一致性问题。由于开发环境、测试环境、生产环境不一致，导致有些 BUG 并未在开发过程中被发现。而 Docker 的镜像提供了除内核外完整的运行时环境，确保了应用运行环境的一致性，从而不会再出现这类问题。

4. 持续交付和部署

对开发和运维（DevOps）人员来说，最希望的就是一次创建或配置就可以在任意地方正常运行。

使用 Docker 可以通过定制应用镜像来实现持续集成、持续交付、部署。开发人员可以通过 Dockerfile 来进行镜像构建，并结合持续集成（Continuous Integration）系统进行集成测试，而运维人员则可以直接在生产环境中快速部署该镜像，甚至结合持续部署(Continuous Delivery/Deployment) 系统进行自动部署。

而且使用 Dockerfile 使镜像构建透明化，不仅使开发团队可以理解应用运行环境，也方便运维团队理解应用运行所需条件，帮助其更好地在生产环境中部署该镜像。

5. 更轻松地迁移

Docker 确保了执行环境的一致性，使得应用的迁移更加容易。Docker 可以在很多平台上运行，无论是物理机、虚拟机、公有云、私有云，甚至是笔记本电脑，其运行结果都是一致的。因此用户可以很轻易地将一个平台上运行的应用迁移到另一个平台上，而不用担心运行环境的变化导致应用无法正常运行的情况。

6. 更轻松地维护和扩展

Docker 使用的分层存储以及镜像技术，使得应用重复部分的复用更为容易，也使得应用的维护更新更加简单，基于基础镜像进一步扩展镜像也变得非常简单。此外，Docker 团队同各个开源项目团队一起维护了一大批高质量的官方镜像，既可以直接在生产环境使用，又可以作为基础进一步加以定制，大大地降低了应用服务的镜像制作成本。

对比容器与传统虚拟机，结果如表 1-1 所示。

表 1-1 容器与虚拟机对比表

特性	容器	虚拟机
启动	秒级	分钟级
磁盘使用	一般为 MB	一般为 GB
性能	接近原生	弱于原生
系统支持量	单机支持上千个容器	一般为几十个容器

1.1.3 Docker 容器的几个概念

1. 容器概念介绍

Docker 的核心概念有以下 3 点。

Docker 容器技术

① 镜像（Image）：一个特殊的文件操作系统，除了提供容器运行时所需的程序、库、资源、配置等文件外，还包含了一些为运行时准备的配置参数（如匿名卷、环境变量、用户等），镜像不包含任何动态数据，其内容在构建之后也不会被改变。

② 容器（Container）：它是用来运行镜像的。例如，拉取了一个 MySQL 镜像之后，只有通过创建并启动 MySQL 容器才能正常地运行 MySQL。容器可以进行创建、启动、停止、删除、暂停等操作。

③ 仓库（Registry）：用来存放镜像文件的地方，可以把自己制作的镜像上传到仓库中，Docker 官方维护了一个公共仓库 Docker Hub。

Docker 的整个运行逻辑如图 1-2 所示。

通过 Docker Client 将需要执行的 Docker 命令发送给 Docker 运行节点上的 Docker daemon，Docker daemon 将请求进行分解执行，例如，执行 docker build 命令时它会根据 Dockerfile 构建一个镜像存放于本地，执行 docker pull 命令时会从远端的容器镜像仓库拉取镜像到本地，执行 docker run 命令时会将容器镜像拉取并运行成为容器实例。

这里可以从容器镜像开始讲解。了解容器镜像先要了解 Linux 系统的基础知识，典型的 Linux 从启动到运行需要 2 个 FS，一个是 rootfs，一个是 bootfs。

bootfs 是 Linux 启动时进行加载的，会安装系统所使用的 kernel 内核,创建完成后 bootfs 会进行解绑。

rootfs 是真正使用 Linux 操作的文件系统，包含/dev/bin 等目录，对于构建容器镜像来说，将 rootfs 作为基础镜像来使用，一个精简的 OS 的 rootfs 可以很小，只需要包括基本的命令和工具即可，例如，alpine 的大小只有不到 10MB。

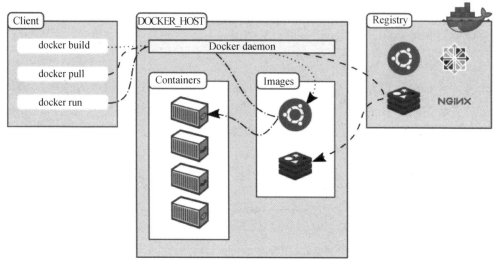

图 1-2　Docker 的整个运行逻辑图

镜像构建基于一个基础镜像将需要进行安装的依赖环境和程序根据 Dockerfile 编写逐层叠加到镜像中。这里涉及了一个容器镜像的概念——镜像分层，镜像的每一层都会有一个独立的 ID，Docker 使用 Union FS 对 Docker 镜像进行分层和合层记录，当使用同一个基础镜像时，利用 Docker 镜像分层的功能可以使用已有的镜像层，拉取没有的镜像层，达到镜像的资源共享，减少重复镜像层拉取，达到资源使用最大化。

容器运行时会在容器镜像最上层生成一个 Container 层，该层复制完整的镜像并加载到内存中运行，在这个层级中对于容器来说是可以进行修改的，但是由于运行于内存中，对 Container 层的任何修改都不会对底层镜像生效，当容器消亡时，修改的数据也一并消亡，这就是容器的 copy on write 特性。

2. Docker 镜像

Docker 镜像类似于虚拟机镜像，可以把它理解为一个只读的模板。例如，一个包含 Nginx 应用程序的镜像，其内部包含一个基本的操作系统环境以及 Nginx 应用程序。镜像是创建 Docker 容器的基础，通过版本管理和增量文件系统，Docker 提供了一套机制来创建或更新现有的镜像，还可以从网上下载并使用别人已经做好的镜像。

简单地理解，Docker 镜像就是一个 Linux 的文件系统，这个文件系统里面包含可以运行在 Linux 内核的程序以及相应的数据。

可以把镜像看成用户空间，当 Docker 通过镜像创建一个容器时，就是将镜像定义好的用户空间作为独立隔离的进程运行在宿主机的 Linux 内核之上。

这里要强调一下镜像的两个特征：

① 镜像是分层（Layer）的，即一个镜像可以由多个中间层组成，多个镜像可以共享同一个中间层，也可以通过在镜像中多添加一层来生成一个新的镜像。

② 镜像是只读的（Read-only），镜像在构建完成之后，便不可以再修改，而上面所说的添加一层构建新的镜像，这实际上是通过创建一个临时的容器，在这个容器上增加或删除文件，从而形成新的镜像，因为容器是可以动态改变的。

如图 1-3 所示，可以帮助我们更好地理解 Docker 镜像与 Linux 的关系。

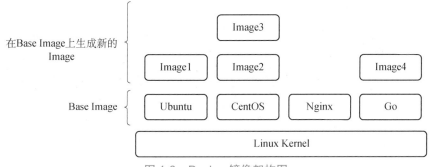

图 1-3 Docker 镜像架构图

3. Docker 容器

Docker 容器类似于一个轻量级沙箱，Docker 利用容器来运行和隔离上层应用。容器与镜像的关系类似于面向对象编程中的对象与类。在镜像中创建的应用在容器中运行实例，从而保证实例之间不会相互影响。容器可以启动、开始、停止、删除，并且这些容器都是彼此相互隔离、互不可见的。

注：镜像本身是只读的，而容器会在启动镜像时，在镜像的最上层创建一个可写层，执行一些必要的写入操作。

因为容器是通过镜像来创建的，所以必须先有镜像才能创建容器，而生成的容器是一个独立于宿主机的隔离进程，并且拥有属于容器自己的网络和命名空间。

前面介绍过，镜像由多个中间层（Layer）组成，生成的镜像是只读的，但容器却是可读可写的，这是因为容器是在镜像上面添一层读写层（Writer/Read Layer）来实现的，如图 1-4 所示。

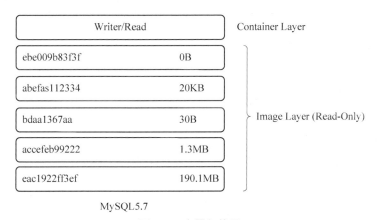

图 1-4 容器架构图

4. Docker 仓库

Docker 仓库类似于代码仓库，是集中存放 Docker 镜像文件的地方。根据所存储的镜像是否公开，可以把仓库分为公开仓库（Public）和私有仓库（Private）两种形式。

目前，最大的公开仓库是 Docker 官方提供的 Docker Hub，其中存放着数量庞大的镜像供用户下载使用。国内也有不少云服务提供商（如腾讯云、阿里云等）提供了仓库的本地源，为用户提供稳定的国内访问。

有时候部门内部有一些镜像要共享时，直接导出镜像拿给别人会比较麻烦，而使用像 Docker Hub 这样的公共仓库又不是很方便，这时候可以自己搭建属于自己的私有仓库服务，用于存储和分布自己的镜像。Docker 官方提供了 Registry 这个镜像，可以用于搭建私有仓库服务。

创建好自有镜像后就可以使用 push 命令将它上传到指定的私有仓库。这样其他人就可以在另外一台机器上下载并使用该镜像，这很像 Git 代码仓库的管理与使用方式。

1.2 Docker 的安装与使用

Docker 引擎的安装（1）

1. 环境准备

（1）规划节点

Docker 节点的规划如表 1-2 所示。

表 1-2　Docker 节点的规划

IP 地址	主机名	节点
192.168.200.10	master	容器节点

（2）基础准备

准备好 VMware Workstation 工具和 CentOS7.5 镜像，创建云主机（云主机配置自定义，推荐配置 2vCPU、4GB 内存、40GB 硬盘，网络使用 NAT 模式，IP 设置为 192.168.200.10，推荐设置主机密码为 000000），使用该虚拟机作为实验基础环境。

注：可以给一个干净的 CentOS7.5 虚拟机创建快照，以备后面克隆使用，可以省去安装的步骤与时间。

2. 安装 Docker 服务

（1）修改主机名

使用远程连接工具连接虚拟机（虚拟机 IP 自行配置为 192.168.200.10），并修改主机名为 master，具体命令如下：

```
[root@localhost ~]# hostnamectl set-hostname master
[root@localhost ~]# logout
[root@master ~]# hostnamectl
   Static hostname: master
         Icon name: computer-vm
           Chassis: vm
        Machine ID: 7378af07fbd948cf8fea45ebcaff9ada
           Boot ID: 8556d6a713434d29802b553c5775cfc8
    Virtualization: vmware
  Operating System: CentOS Linux 7 (Core)
       CPE OS Name: cpe:/o:centos:centos:7
```

```
          Kernel: Linux 3.10.0-862.el7.x86_64
        Architecture: x86-64
```

(2)关闭防火墙及 SELinux

使用如下命令,关闭防火墙及 SELinux 服务:

```
[root@master ~]# setenforce 0
[root@master ~]# getenforce
Permissive
[root@master ~]# systemctl stop firewalld
```

(3)配置 YUM 源

将提供的 docker-repo 目录上传至虚拟机的/opt 目录下,并配置本地 YUM 源文件 local.repo,具体步骤如下。

首先移除/etc/yum.repos.d/目录下的原有文件到/media 目录,命令如下:

```
[root@master ~]# mv /etc/yum.repos.d/* /media/
```

创建新的 local.repo 文件,命令与文件内容如下:

```
[root@master ~]# vi /etc/yum.repos.d/local.repo
[root@master ~]# cat /etc/yum.repos.d/local.repo
[docker]
name=docker
baseurl=file:///opt/docker-repo
gpgcheck=0
enabled=1
```

查看配置的本地 YUM 源是否可用,命令如下:

```
[root@master ~]# yum clean all
Loaded plugins: fastestmirror
Cleaning repos: docker
Cleaning up everything
Maybe you want: rm -rf /var/cache/yum, to also free up space taken by orphaned data from disabled or removed repos
[root@master ~]# yum repolist
Loaded plugins: fastestmirror
Determining fastest mirrors
docker                          | 2.9 kB  00:00:00
docker/primary_db               | 166 kB  00:00:00
repo id            repo name                    status
docker             docker                       178
repolist: 178
```

可以看到 repolist 的数量为 178,即 YUM 源可用。

（4）安装 Docker 服务

使用 yum 命令安装 docker-ce 服务，命令如下：

```
[root@master ~]# yum install docker-ce -y
... ...
Installed:
  docker-ce.x86_64 3:19.03.13-3.el7
... ...
Complete!
```

Docker 引擎的安装（2）

安装 Docker 服务完毕，启动并查看 Docker 服务版本，命令如下：

```
[root@master ~]# systemctl start docker
[root@master ~]# docker info
Client:
 Debug Mode: false

Server:
 Containers: 0
  Running: 0
  Paused: 0
  Stopped: 0
 Images: 0
 Server Version: 19.03.13
 Storage Driver: overlay2
  Backing Filesystem: xfs
  Supports d_type: true
  Native Overlay Diff: true
 Logging Driver: json-file
 Cgroup Driver: cgroupfs
 Plugins:
  Volume: local
  Network: bridge host ipvlan macvlan null overlay
  Log: awslogs fluentd gcplogs gelf journald json-file local logentries splunk syslog
 Swarm: inactive
 Runtimes: runc
 Default Runtime: runc
 Init Binary: docker-init
 containerd version: 8fba4e9a7d01810a393d5d25a3621dc101981175
 runc version: dc9208a3303feef5b3839f4323d9beb36df0a9dd
 init version: fec3683
 Security Options:
  seccomp
   Profile: default
```

```
Kernel Version: 3.10.0-862.el7.x86_64
Operating System: CentOS Linux 7 (Core)
OSType: linux
Architecture: x86_64
CPUs: 2
Total Memory: 3.685GiB
Name: master
ID: BSHC:YAO7:DHUQ:N5Y3:RKFI:3G7D:5IZI:TNTP:PHU5:J3DA:VPFE:4DJ3
Docker Root Dir: /var/lib/docker
Debug Mode: false
Registry: https://index.docker.io/v1/
Labels:
Experimental: false
Insecure Registries:
 127.0.0.0/8
Live Restore Enabled: false
```

可以看到 Docker 的版本为 19.03.13。

3. Docker 基本命令

（1）镜像基础命令

Docker 中与镜像操作相关的命令都在 docker image 这条子命令下，通过 docker image--help 命令，可以看到 docker image 子命令的详细文档，内容如下：

```
Usage:  docker image COMMAND

Manage images

Commands:
  build     Build an image from a Dockerfile(构建镜像的命令)
  history   Show the history of an image(显示镜像构建历史过程)
  import    Import the contents from a tarball to create a filesystem image(导入一个由容器导出的镜像)
  inspect   Display detailed information on one or more images(显示一个镜像的详细信息)
  load      Load an image from a tar archive or STDIN(从一个文件或标准输入流中导入镜像)
  ls        List images(查看镜像列表)
  prune     Remove unused images(删除虚悬镜像)
  pull      Pull an image or a repository from a registry(从仓库拉取镜像)
  push      Push an image or a repository to a registry(推送镜像到仓库)
  rm        Remove one or more images(删除镜像)
  save    Save one or more images to a tar archive (streamed to STDOUT by default)(保存镜像到文件)
```

```
  tag         Create a tag TARGET_IMAGE that refers to SOURCE_IMAGE(给镜像打
标签)
```

获取镜像：在安装了 Docker 之后，本地并没有任何镜像，当然可以自己构建，不过更方便的还是从 Docker 官方提供的仓库服务 Docker Hub 上拉取官方或第三方已经构建好的镜像。

拉取镜像可以使用 docker image pull 命令，其格式如下：

```
docker image pull [OPTIONS] NAME[:TAG|@DIGEST]
```

当然，docker image pull 有更简洁的用法，例如：

```
docker pull [OPTIONS] NAME[:TAG|@DIGEST]
```

要获取镜像，需要指定 Docker Registry 的 URL 和端口号，默认是 Docker Hub，另外还需要指定仓库名和标签，仓库名和标签可以唯一确定一个镜像，而标签是可以省略的，如果省略，则默认使用 latest 作为标签名，而仓库名则由作者名和软件名组成。

所以，在省略了参数后，比如拉取 centos 镜像时，就可以使用下面简单的命令从 Docker Hub 上拉取到（拉取镜像时要确保虚拟机能访问外网）：

```
[root@master ~]# docker pull centos
Using default tag: latest
latest: Pulling from library/centos
7a0437f04f83: Pull complete
Digest: sha256:5528e8b1b1719d34604c87e11dcd1c0a20bedf46e83b5632cdeac91b8c04efc1
Status: Downloaded newer image for centos:latest
docker.io/library/centos:latest
```

查看本地镜像：通过上面的方法将镜像拉取到了本地后，那要如何查看本地有哪些镜像呢？通过下面的命令可以查看本地的全部镜像：

```
[root@master ~]# docker image ls
REPOSITORY       TAG          IMAGE ID         CREATED          SIZE
centos           latest       300e315adb2f     7 weeks ago      209MB
```

当然 Docker 提供了更简洁的写法，例如：

```
[root@master ~]# docker images
REPOSITORY       TAG          IMAGE ID         CREATED          SIZE
centos           latest       300e315adb2f     7 weeks ago      209MB
```

Docker 镜像名由仓库名和标签组成，但有时候会看到仓库名和标签皆为<none>的镜像，我们称这种镜像为虚悬镜像，命令如下：

```
[root@master ~]# docker images
REPOSITORY       TAG          IMAGE ID         CREATED          SIZE
centos           latest       300e315adb2f     7 weeks ago      209MB
<none>           <none>       6858809bf669     4 months ago     1.23MB
```

将提供的 busybox_latest.tar 上传至虚拟机的/root 目录下，并将镜像导入，命令如下：

```
[root@master ~]# docker load -i busybox_latest.tar
be8b8b42328a: Loading layer [====================>]   1.45MB/1.45MB
Loaded image ID: sha256:6858809bf669cc5da7cb6af83d0fae838284d12e1be0182f92f6bd96559873e3
```

导入该镜像之后，使用 docker images 命令查看镜像列表，即能看到虚悬镜像。

```
[root@master ~]# docker images
REPOSITORY       TAG           IMAGE ID        CREATED         SIZE
centos           latest        300e315adb2f    7 weeks ago     209MB
<none>           <none>        6858809bf669    4 months ago    1.23MB
```

虚悬镜像一般是用户使用 docker pull 拉取最新镜像时生成的，此时仓库名和标签给了新的镜像，旧的镜像仓库和标签则被取消，成为虚悬镜像。

用户可以使用下面的语句打印所有的虚悬镜像，命令如下：

```
[root@master ~]# docker images -f dangling=true
REPOSITORY       TAG           IMAGE ID        CREATED         SIZE
<none>           <none>        6858809bf669    4 months ago    1.23MB
```

一般来说，虚悬镜像已经失去了存在的价值，是可以随意删除的，可以用下面的命令删除。

```
[root@master ~]# docker image prune
WARNING! This will remove all dangling images.
Are you sure you want to continue? [y/N] y
Deleted Images:
deleted: sha256:6858809bf669cc5da7cb6af83d0fae838284d12e1be0182f92f6bd96559873e3
deleted: sha256:be8b8b42328a15af9dd6af4cba85821aad30adde28d249d1ea03c74690530d1c

Total reclaimed space: 1.232MB
[root@master ~]# docker images
REPOSITORY       TAG           IMAGE ID        CREATED         SIZE
centos           latest        300e315adb2f    7 weeks ago     209MB
```

如果想与别人共享某个镜像，除了从镜像服务仓库中 pull 镜像和把镜像 push 到仓库上之外，其实还可以将本地构建好的镜像直接导出并保存为文件发送给别人。例如，当前有一个 centos 镜像，可以使用命令保存到本地，命令如下：

```
[root@master ~]# docker images
REPOSITORY       TAG           IMAGE ID        CREATED         SIZE
centos           latest        300e315adb2f    7 weeks ago     209MB
[root@master ~]# docker save centos:latest > centos_latest.tar
```

```
[root@master ~]# ll
total 212888
-rw-------.  1 root root      1572 Aug 10 09:42 anaconda-ks.cfg
-rw-r--r--.  1 root root   1458176 Jan 26 00:48 busybox_latest.tar
-rw-r--r--.  1 root root 216535040 Jan 26 01:48 centos_latest.tar
```

而当用户拿到别人导出的镜像文件时，可以使用 docker load 命令把镜像加载到本地的 Docker 镜像列表中。例如，要加载 busybox_latest.tar 镜像，命令如下：

```
[root@master ~]# docker load -i busybox_latest.tar
be8b8b42328a: Loading layer [==================>]  1.45MB/1.45MB
Loaded image ID: sha256:6858809bf669cc5da7cb6af83d0fae838284d12e1be0182f92f6bd96559873e3
[root@master ~]# docker images
REPOSITORY     TAG        IMAGE ID        CREATED         SIZE
centos         latest     300e315adb2f    7 weeks ago     209MB
<none>         <none>     6858809bf669    4 months ago    1.23MB
```

新加载的 busybox_latest.tar 镜像会变成虚悬镜像，如果想保留一些有用的虚悬镜像，则可以使用 docker tag 命令重新给镜像起仓库名和标签，命令如下：

```
[root@master ~]# docker tag 6858809bf669 busybox:latest
[root@master ~]# docker images
REPOSITORY     TAG        IMAGE ID        CREATED         SIZE
centos         latest     300e315adb2f    7 weeks ago     209MB
busybox        latest     6858809bf669    4 months ago    1.23MB
```

要删除一个或多个本地的镜像时，可以使用下面的命令：

```
docker image rm [option] IMAGE1,IMAGE2,...IMAGEn
```

也可以使用更简洁的方式，如：

```
docker rmi [option] IMAGE1,IMAGE2,...IMAGEn
```

可以使用镜像长 ID、镜像短 ID、镜像摘要以及镜像名称来删除镜像，一般会使用镜像 ID 来删除镜像，具体命令如下（删除 busybox 镜像）：

```
[root@master ~]# docker images
REPOSITORY     TAG        IMAGE ID        CREATED         SIZE
centos         latest     300e315adb2f    7 weeks ago     209MB
busybox        latest     6858809bf669    4 months ago    1.23MB
[root@master ~]# docker rmi 6858809bf669
Untagged: busybox:latest
Deleted: sha256:6858809bf669cc5da7cb6af83d0fae838284d12e1be0182f92f6bd96559873e3
Deleted: sha256:be8b8b42328a15af9dd6af4cba85821aad30adde28d249d1ea03c74690530d1c
```

```
[root@master ~]# docker images
REPOSITORY      TAG        IMAGE ID         CREATED        SIZE
centos          latest     300e315adb2f     7 weeks ago    209MB
```

以上就是关于 Docker 镜像服务经常使用的一些命令，接下来学习容器基础使用命令。

（2）容器基础命令

操作容器的相关命令参数如下：

```
Usage: docker container COMMAND

Manage containers

Commands:
    attach      Attach local standard input, output, and error streams to a running container
    commit      Create a new image from a container's changes(把容器保存为镜像)
    cp          Copy files/folders between a container and the local filesystem
    create      Create a new container(创建一个新的容器)
    diff        Inspect changes to files or directories on a container's filesystem
    exec        Run a command in a running container(在一个运行的容器中执行命令)
    export      Export a container's filesystem as a tar archive
    inspect     Display detailed information on one or more containers
    kill        Kill one or more running containers(杀死一个或多个正在运行的容器)
    logs        Fetch the logs of a container
    ls          List containers(显示本地容器列表)
    pause       Pause all processes within one or more containers
    port        List port mappings or a specific mapping for the container
    prune       Remove all stopped containers
    rename      Rename a container(重命名容器)
    restart     Restart one or more containers(重启一个或多个容器)
    rm          Remove one or more containers(删除一个或多个容器)
    run         Run a command in a new container(运行一个新的容器)
    start       Start one or more stopped containers
    stats       Display a live stream of container(s) resource usage statistics
    stop        Stop one or more running containers(停止一个或多个容器)
    top         Display the running processes of a container
    unpause     Unpause all processes within one or more containers
    update      Update configuration of one or more containers
    wait        Block until one or more containers stop, then print their exit codes
```

启动容器：启动容器有几种不同的方式，最常用的方法是使用 docker run 命令，可以通过镜像创建一个容器，命令如下：

```
# /bin/bash 表示运行容器后要执行的命令；
```

```
# -it:以交互模式运行容器,为容器重新分配一个伪输入终端,通常一起使用;
$ docker run -it centos /bin/bash
```

docker run 命令有一些比较常用的参数,比如容器是一种提供服务的守护进程,那么通常需要开放端口供外部访问,命令如下。

首先将提供的 nginx_latest.tar 上传至虚拟机的/root 目录下,使用 load 命令加载至本地:

```
[root@master ~]# docker load -i nginx_latest.tar
07cab4339852: Loading layer [==================>]  72.49MB/72.49MB
822ae9fef1d8: Loading layer [==================>]  64.53MB/64.53MB
7230cfe05cc1: Loading layer [==================>]  3.072kB/3.072kB
8eb80f066de2: Loading layer [==================>]  4.096kB/4.096kB
8032102adebe: Loading layer [==================>]  3.584kB/3.584kB
Loaded image: nginx:latest
```

使用 Nginx 镜像,启动容器,并映射内部的 80 端口到外部的 80 端口,具体命令如下:

```
[root@master ~]# docker run -p 80:80 nginx
/docker-entrypoint.sh: /docker-entrypoint.d/ is not empty, will attempt to perform configuration
/docker-entrypoint.sh: Looking for shell scripts in /docker-entrypoint.d/
/docker-entrypoint.sh: Launching /docker-entrypoint.d/10-listen-on-ipv6-by-default.sh
10-listen-on-ipv6-by-default.sh: Getting the checksum of /etc/nginx/conf.d/default.conf
10-listen-on-ipv6-by-default.sh: Enabled listen on IPv6 in /etc/nginx/conf.d/default.conf
/docker-entrypoint.sh: Launching /docker-entrypoint.d/20-envsubst-on-templates.sh
/docker-entrypoint.sh: Configuration complete; ready for start up
```

可以通过访问网页 http://192.168.200.10/查看 Nginx 服务的首页,如图 1-5 所示。

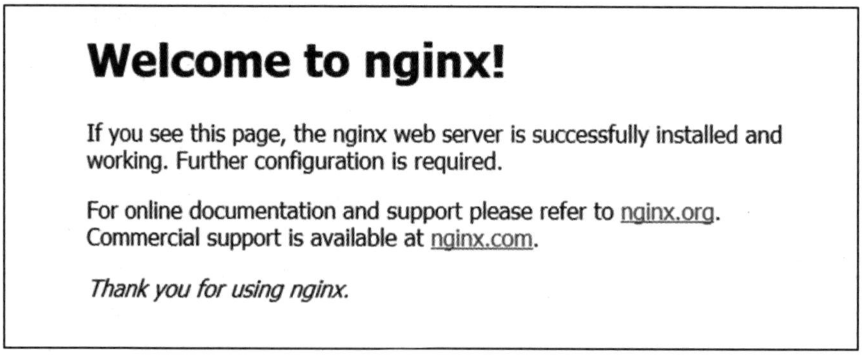

图 1-5　Nginx 服务的首页

但是此时命令行界面会被该容器进程占据,一旦使用 Ctrl+C 键退出之后,Nginx 服务也会随之被关闭,所以 Docker 提供了-d 参数,可以让容器在后台运行,命令如下:

```
[root@master ~]# docker run -d -p 80:80 nginx
197de6e73f69c7378f4fbaa847b54ed650b6043290f6022567a8353417658555
```

也可以为容器指定一个名字，具体命令如下（此处不能使用 80 端口，因为在上一步操作中，宿主机的 80 端口已被使用）：

```
[root@master ~]# docker run -d -p 81:80 --name webserver nginx
bf03da1f1d68df6c2f61d1899f81bf81933fd7bbacc0402d3649eab5fe90a54c
```

可以使用 docker stop 命令停止正在运行的容器，将 webserver 容器停止。

首先使用 docker ps -a 命令查询容器列表，命令如下：

```
[root@master ~]# docker ps -a
  CONTAINER ID        IMAGE               COMMAND                  CREATED
STATUS                  PORTS               NAMES
  bf03da1f1d68        nginx               "/docker-entrypoint.…"   6 minutes ago
Up 6 minutes            0.0.0.0:81->80/tcp  webserver
  197de6e73f69        nginx               "/docker-entrypoint.…"   38 minutes
ago    Up 38 minutes    0.0.0.0:80->80/tcp  cranky_pasteur
  8ba36f089de0        nginx               "/docker-entrypoint.…"   57 minutes
ago    Exited (0) 54 minutes ago                priceless_boyd
```

然后使用 docker stop 命令停止正在运行的 webserver 容器，命令如下：

```
[root@master ~]# docker stop bf03da1f1d68
bf03da1f1d68
```

继续用 docker ps -a 命令查看容器列表，可以看到 webserver 容器处于 Exited 状态，命令如下：

```
[root@master ~]# docker ps -a
  CONTAINER ID        IMAGE               COMMAND                  CREATED
STATUS                  PORTS               NAMES
  bf03da1f1d68        nginx               "/docker-entrypoint.…"   9 minutes ago
Exited (0) About a minute ago                webserver
  197de6e73f69        nginx               "/docker-entrypoint.…"   41 minutes
ago    Up 41 minutes    0.0.0.0:80->80/tcp  cranky_pasteur
  8ba36f089de0        nginx               "/docker-entrypoint.…"   About an hour
ago    Exited (0) 57 minutes ago                priceless_boyd
```

使用 docker start 命令重新启动已经停止运行的 webserver 容器，具体命令如下：

```
[root@master ~]# docker start bf03da1f1d68
bf03da1f1d68
```

而对于正在运行的容器，也可以通过 docker restart 命令重新启动，命令如下：

```
[root@master ~]# docker restart bf03da1f1d68
bf03da1f1d68
```

Docker 容器大体上有 3 种运行模式，具体如下。

① 运行后退出。使用如下命令创建的容器，在运行后会退出：

```
[root@master ~]# docker run centos echo "hellowrold"
hellowrold
```

② 守护进程模式。如果容器中运行一个守护进程，则容器会一直处于运行状态，具体命令如下：

```
[root@master ~]# docker run -d -p 82:80 nginx
a3339b25e46036ca8b930572d0083250985930781cf4f90b96f804deba0296c3
```

③ 交互式。用户也可以在运行容器时，直接与容器进行交互，具体命令如下：

```
[root@master ~]# docker run -it centos /bin/bash
[root@7dfe3a1ca51f /]#
```

可以使用 exit 命令退出或者按 Ctrl+D 键退出交互界面。

运行容器后，用户可以通过下面的命令查看本地所有容器列表，命令如下：

```
[root@master ~]# docker ps -a
CONTAINER ID      IMAGE           COMMAND                  CREATED        STATUS              PORTS                  NAMES
bf03da1f1d68      nginx           "/docker-entrypoint.…"   17 minutes ago    Up 4 minutes        0.0.0.0:81->80/tcp     webserver
197de6e73f69      nginx           "/docker-entrypoint.…"   49 minutes ago    Up 49 minutes       0.0.0.0:80->80/tcp     cranky_pasteur
8ba36f089de0      nginx           "/docker-entrypoint.…"   About an hour ago    Exited (0) About an hour ago              priceless_boyd
```

若只想查找容器的 ID，可以使用下面的命令：

```
[root@master ~]# docker ps -aq
bf03da1f1d68
197de6e73f69
8ba36f089de0
```

删除容器：删除容器的命令很简洁，命令如下：

```
[root@master ~]# docker ps -a
CONTAINER ID      IMAGE           COMMAND                  CREATED        STATUS              PORTS                  NAMES
206754da4743      centos          "/bin/bash"              2 minutes ago    Exited (0) About a minute ago                 keen_hamilton
a3339b25e460      nginx           "/docker-entrypoint.…"   3 minutes ago    Up 3 minutes        0.0.0.0:82->80/tcp     vibrant_sinoussi
bf03da1f1d68      nginx           "/docker-entrypoint.…"   26 minutes ago    Up 13 minutes       0.0.0.0:81->80/tcp     webserver
197de6e73f69      nginx           "/docker-entrypoint.…"   58 minutes
```

```
ago       Up 58 minutes                 0.0.0.0:80->80/tcp   cranky_pasteur
   8ba36f089de0        nginx              "/docker-entrypoint.…"   About an hour
ago    Exited (0) About an hour ago                   priceless_boyd
   [root@master ~]# docker rm 206754da4743
   206754da4743
```

如果想删除正在运行的容器，则需要加-f 参数，具体命令如下：

```
[root@master ~]# docker ps -a
   CONTAINER ID        IMAGE              COMMAND                  CREATED
STATUS              PORTS              NAMES
   a3339b25e460        nginx              "/docker-entrypoint.…"   4 minutes ago
Up 4 minutes           0.0.0.0:82->80/tcp   vibrant_sinoussi
   bf03da1f1d68        nginx              "/docker-entrypoint.…"   27 minutes
ago    Up 15 minutes              0.0.0.0:81->80/tcp   webserver
   197de6e73f69        nginx              "/docker-entrypoint.…"   59 minutes
ago    Up 59 minutes              0.0.0.0:80->80/tcp   cranky_pasteur
   8ba36f089de0        nginx              "/docker-entrypoint.…"   About an hour
ago    Exited (0) About an hour ago                   priceless_boyd
   [root@master ~]# docker rm -f a3339b25e460
   a3339b25e460
```

也可以批量删除已经停止运行的容器，正在运行的容器不会被删除，但是会报出一个不能删除的错误，命令如下：

```
[root@master ~]# docker rm $(docker ps -aq)
   8ba36f089de0
   Error response from daemon: You cannot remove a running container bf03da1f1d68df6c2f61d1899f81bf81933fd7bbacc0402d3649eab5fe90a54c. Stop the container before attempting removal or force remove
   Error response from daemon: You cannot remove a running container 197de6e73f69c7378f4fbaa847b54ed650b6043290f6022567a8353417658555. Stop the container before attempting removal or force remove
```

进入容器：对于正在运行的容器，用户也可以通过 docker exec 命令再次进入容器，命令如下：

```
[root@master ~]# docker ps -a
   CONTAINER ID        IMAGE              COMMAND                  CREATED
STATUS              PORTS              NAMES
   bf03da1f1d68        nginx              "/docker-entrypoint.…"   30 minutes
ago    Up 17 minutes       0.0.0.0:81->80/tcp   webserver
   197de6e73f69        nginx              "/docker-entrypoint.…"   About an hour
ago    Up About an hour    0.0.0.0:80->80/tcp   cranky_pasteur
   [root@master ~]# docker exec -it bf03da1f1d68 /bin/bash
   root@bf03da1f1d68:/#
```

```
root@bf03da1f1d68:/# ls
bin   dev   docker-entrypoint.sh  home  lib64  mnt    proc  run   srv   tmp  var
boot  docker-entrypoint.d   etc         lib    media  opt   root  sbin  sys  usr
```

在容器中，一些基础的命令，如 ls 等，也可以被用来操作容器，读者可以自行尝试。

归纳总结

通过本单元内容的学习，读者应该对 Docker 容器服务有了一定的认识，也熟悉了容器中的几个重要概念。通过实操练习，掌握了 Docker 容器服务的安装与基本使用命令。但是关于 Docker 容器的使用，远不止这么简单，在后面的课程中，会进一步学习容器仓库、网络、镜像、存储等内容。

课后练习

一、判断题

1. Docker 使用 Docker Hub 平台来管理公开的镜像。（   ）
2. 镜像丰富是 Docker 官方镜像仓库的一大优点。（   ）

二、单项选择题

1. 在下列选项当中，容器提供了（   ）平台服务。

   A. IaaS   B. PaaS   C. SaaS   D. CaaS

2. 下面（   ）命令可以从镜像仓库中获取一个 Docker 镜像。

   A. docker get   B. docker pull   C. docker put   D. docker push

三、多项选择题

1. Docker 容器有（   ）状态。

   A. 运行   B. 已暂停   C. 重新启动   D. 已退出

2. Docker 的核心组件有（   ）。

   A. 镜像   B. 容器   C. 仓库   D. 组件

技能训练

1. 安装 Docker 容器服务，并查看容器的版本号。

2. 启动 Registry 容器时，将内部保存文件的目录映射到外部的/opt/registry 目录，将内部的 5000 端口映射到外部 5000 端口。

单元 2　Docker 镜像仓库

学习目标

通过本单元的学习,要求了解 Docker 镜像仓库的概念、Docker 镜像仓库的部署和使用,了解 Registry 仓库的功能及特点,了解 Harbor 的优势及仓库架构等。本单元培养读者掌握 Registry 镜像仓库的安装与使用、Harbor 镜像仓库的安装和使用、使用命令对仓库进行运维等技能;让读者掌握相关基础理论知识和动手实操的能力。

2.1　Docker 仓库的基本概念

2.1.1　Registry 私有镜像仓库

Docker 仓库的使用与维护(1)

1. Registry 简介

官方的 Docker Hub 是一个用于管理公共镜像的好地方,用户可以在上面找到想要的镜像,也可以把自己创建的镜像推送上去。但是有时候,容器的使用场景需要用户拥有一个私有的镜像仓库以用于管理自己的镜像。这个可以通过开源软件 Registry 来达成目的。

Registry 在 Github 上有两份代码:老代码库和新代码库。老代码库是采用 Python 编写的,存在 pull 和 push 的性能问题,0.9.1 版本之后就标志为 deprecated,不再继续开发。新代码库是采用 Go 语言编写的,修改了镜像 ID 的生成算法、Registry 上镜像的保存结构,大大优化了 pull 和 push 镜像的效率。

官方在 Docker Hub 上提供了 Registry 的镜像,用户可以直接使用该 Registry 镜像来构建一个容器,搭建属于自己的私有仓库服务。

2. Registry 的功能

(1)镜像存储

镜像数据存储在 Registry 后端,与本地镜像存储方式类似,它也分隔了多个镜像层,放置在各自的目录中,保存成 tar 包格式。除了镜像数据外,Registry 还保存了清单文件(manifest)和镜像签名文件(signature)等。

(2)镜像创建、分发和更新

本地用户和 Registry 之间通过 Registry API 传输镜像。Registry API 即一系列 HTTP/HTTPS 请求,用来发送用户请求到 Registry,并接收 Registry 的响应。在请求和响应中则包含了镜像数据交互。

3. Registry 的特点

Registry 镜像仓库是一个非常简单易用的私有镜像仓库服务，其主要的特点如下：
① 快速上传和下载镜像。
② 部署方便。
③ 后端支持多种分布式云存储方案（如 S3、Azure）和本地文件系统等，接口以插件方式存在，易于配置。
④ 拥有完善的镜像缓存机制，镜像下载更加快捷。
⑤ 以 Webhook 方式实现通知系统。
⑥ 实现了本地 TLS，支持 HTTPS 安全访问。
⑦ 有可配置的认证模块。
⑧ 有健康检查模块。
⑨ 正在努力让用于管理镜像的清单和摘要文件格式更加清晰，以及更清楚地为镜像打标签。
⑩ 设计方案新颖且高性能。
⑪ 清单文件（manifest）作为元数据完整地记录镜像信息。

2.1.2　Harbor 企业级镜像仓库

1. Harbor 镜像仓库简介

Harbor 是由 VMware 公司开源的企业级的 Docker Registry 管理项目，Harbor 主要提供 Docker Registry 管理 UI，提供的功能包括：基于角色访问的控制权限管理（RBAC）、AD/LDAP 集成、日志审核、管理界面、自我注册、镜像复制和中文支持等。Harbor 的目标是帮助用户迅速搭建一个企业级的 Docker Registry 服务。它以 Docker 公司开源的 Registry 为基础。

Harbor 除了提供友好的 Web UI 界面、角色和用户权限管理、用户操作审计等功能外，它还整合了 k8s 的插件（Add-ons）仓库，即 Helm 通过 chart 方式下载、管理、安装 k8s 插件，而 chartmuseum 可以提供存储 chart 数据的仓库。

注：helm 就相当于 k8s 的 YUM。

另外它还整合了两个开源的安全组件，一个是 Notary，另一个是 Clair。Notary 类似于私有 CA 中心，而 Clair 则是容器安全扫描工具，它通过各大厂商提供的 CVE 漏洞库来获取最新漏洞信息，并扫描用户上传的容器中是否存在已知的漏洞信息，这两个安全功能对于企业级私有仓库来说是非常有意义的。

2. Harbor 的优势

Harbor 镜像仓库的优势如下。
① 基于角色控制：角色有管理员与普通用户，可赋权普通用户，比如只能上传和下载，可根据项目来进行操作和管理。

② 基于镜像的复制策略：它也与权限相关，比如只有特有的用户与组才能对此项目进行相对应的操作。

③ 支持 LDAP/AD：域控制，比如在南京去下载北京 Harbor 私有仓库的镜像，两端输入局域网的地址，连接在一块，数据信息的传输通过一条隧道，会通过两层加密，第一层为隧道加密，第二层为数据加密，安全可靠。

④ 图像删除和垃圾收集：采用回收站机制。

⑤ 图形 UI：具有统计功能，比如访问量与镜像下载热度。

⑥ 审计：日志，在这里意义不大，主要借助于 ELK。

⑦ RESTful API：定义 Web 语言规范的格式，方便调用 Harbor 的接口，也便于二次开发。

3. Harbor 仓库架构

Harbor 的每个组件都是以 Docker 容器的形式构建的，可以使用 Docker Compose 来进行部署。如果环境中使用了 Kubernetes，Harbor 也提供了 Kubernetes 的配置文件。Harbor 大概需要以下几个容器组成：

① UI（Harbor 的核心服务）。

② Log（运行着 rsyslog 的容器，进行日志收集）。

③ MySQL（由官方 MySQL 镜像构成的数据库容器）。

④ Nginx（使用 Nginx 做反向代理）。

⑤ Registry（官方的 Docker Registry）。

⑥ Adminserver（Harbor 的配置数据管理器）。

⑦ Jobservice（Harbor 的任务管理服务）。

⑧ Redis（用于存储 session）。

2.2 Docker 镜像仓库的部署和使用

Docker 仓库的使用与维护（2）

1. 环境准备

（1）规划节点

仓库节点的规划如表 2-1 所示。

表 2-1 仓库节点的规划

IP 地址	主机名	节点
192.168.200.11	master	Docker 仓库节点
192.168.200.12	client	Docker 客户端节点

（2）基础准备

准备 VMware Workstation 工具，使用 CentOS7.5 虚拟机快照，克隆两台虚拟机，云主机配置自定义，推荐配置 2vCPU、4GB 内存、40GB 硬盘，网络使用 NAT 模式，master 节点 IP 设置为 192.168.200.11；client 节点 IP 设置为 192.168.200.12，推荐设置主机密码为 000000，使用克隆的两台虚拟机作为实验基础环境。在进行仓库实验前，自行将两台虚拟机配置好 IP、主机名并安装 Docker 服务，关闭 SELinux 和防火墙服务（安装 Docker 服务的实验步骤可参考 1.2 节）。

2. Registry 镜像仓库的安装与使用

（1）Registry 安装

登录到 master 节点，该节点已安装 Docker 服务，使用 docker pull 命令拉取 Registry 镜像，具体命令如下：

```
[root@master ~]# docker pull registry
Using default tag: latest
latest: Pulling from library/registry
0a6724ff3fcd: Pull complete
d550a247d74f: Pull complete
1a938458ca36: Pull complete
acd758c36fc9: Pull complete
9af6d68b484a: Pull complete
Digest: sha256:d5459fcb27aecc752520df4b492b08358a1912fcdfa454f7d2101d4b09991daa
Status: Downloaded newer image for registry:latest
docker.io/library/registry:latest
[root@master ~]# docker images
REPOSITORY     TAG       IMAGE ID       CREATED       SIZE
registry       latest    678dfa38fcfa   5 weeks ago   26.2MB
```

可以看到 Registry 镜像已经被拉取下来，因为在 pull 的时候没有指定镜像的版本，所以默认拉取的版本是最新（latest）版本。使用如下命令部署私有仓库：

```
[root@master ~]# docker run -d -p 5000:5000 --restart=always --name registry registry:latest
584a017941fe642d0a9ab17a0ff19f52650b5612a5a0cc5412d097950b0a1057
```

该命令中各参数的意义如下：
- #：创建一个 Registry 容器来运行 Registry 服务。
- -p：端口映射（前面是宿主机端口，后面是容器暴露的端口）。
- --restart=always：随 Docker 服务的启动而启动。
- --name registry：指定容器名为 registry。

使用 docker ps -a 命令查看容器列表：

```
[root@master ~]# docker ps -a
CONTAINER ID    IMAGE    COMMAND    CREATED    STATUS    PORTS    NAMES
```

```
584a017941fe        registry: latest       "/entrypoint.sh /etc…"   25 seconds
ago       Up 24 seconds       0.0.0.0: 5000->5000/tcp    registry
```

启动 Registry 容器后,打开浏览器输入 http://ip_addr:5000/v2/(此处的 ip.addr 为 192.168.200.11),看到如图 2-1 所示的界面说明 Registry 运行正常。

图 2-1　Registry 界面

（2）测试镜像上传

将镜像 push 到 Registry 来验证,首先将提供的 nginx_latest.tar 镜像上传至 master 节点的/root 目录下,然后使用 docker load 命令加载至本地,命令如下：

```
[root@master ~]# docker load -i nginx_latest.tar
07cab4339852: Loading layer [==============>]  72.49MB/72.49MB
822ae9fef1d8: Loading layer [==============>]  64.53MB/64.53MB
7230cfe05cc1: Loading layer [==============>]  3.072kB/3.072kB
8eb80f066de2: Loading layer [==============>]  4.096kB/4.096kB
8032102adebe: Loading layer [==============>]  3.584kB/3.584kB
Loaded image: nginx:latest
[root@master ~]# docker image
REPOSITORY        TAG         IMAGE ID         CREATED         SIZE
registry          latest      678dfa38fcfa     5 weeks ago     26.2MB
nginx             latest      992e3b7be046     3 months ago    133MB
```

将想要上传到镜像仓库的镜像命名为符合仓库要求 registry_url:port/ImageName:tag 的格式,可以通过 docker tag 命令来实现,具体命令如下：

```
[root@master ~]# docker tag 992e3b7be046 192.168.200.11:5000/nginx:latest
[root@master ~]# docker images
REPOSITORY                         TAG       IMAGE ID         CREATED         SIZE
registry                           latest    678dfa38fcfa     5 weeks ago     26.2MB
nginx                              latest    992e3b7be046     3 months ago    133MB
192.168.200.11:5000/nginx          latest    992e3b7be046     3 months ago    133MB
```

上传 Nginx 镜像到 Registry 镜像仓库,具体命令如下：

```
[root@master ~]# docker push 192.168.200.11:5000/nginx:latest
The push refers to repository [192.168.200.11:5000/nginx]
Get https://192.168.200.11:5000/v2/: dial tcp 192.168.200.11:5000: connect:
no route to host
```

在上传的过程中遇到了报错，这是因为启动的 Registry 服务不是安全可信赖的。这时需要修改客户端 Docker 的配置文件/etc/docker/daemon.json（若该文件不存在则创建一个），修改成如下命令：

```
[root@master ~]# cat /etc/docker/daemon.json
{
"insecure-registries": ["192.168.200.11:5000"]
}
```

修改完毕后，重启 Docker 服务并上传镜像至仓库，命令如下：

```
[root@master ~]# systemctl restart docker
[root@master ~]# docker push 192.168.200.11:5000/nginx:latest
The push refers to repository [192.168.200.11:5000/nginx]
8032102adebe: Pushed
8eb80f066de2: Pushed
7230cfe05cc1: Pushed
822ae9fef1d8: Pushed
07cab4339852: Pushed
latest: digest: sha256:416d511ffa63777489af47f250b70d1570e428b67666567085f2bece3571ad83 size: 1362
```

push 成功之后，可以在浏览器中输入 http://ip_add:5000/v2/_catalog 查看镜像是否被推送到镜像仓库（此外 ip_add 为 192.168.200.11），如图 2-2 所示。

图 2-2　Registry 界面

或者使用以下命令查看：

```
[root@master ~]# curl http://192.168.200.11:5000/v2/_catalog
{"repositories":["nginx"]}
[root@master ~]# curl http://192.168.200.11:5000/v2/nginx/tags/list
{"name":"nginx","tags":["latest"]}
```

至此，Registry 私有仓库搭建完成，接下来使用一个客户端节点进行私有仓库中镜像的拉取。

（3）客户端拉取镜像

使用远程连接工具连接到 client 节点，client 节点已经安装了 Docker 服务并启动。

首先需要修改 client 节点 Docker 服务的配置文件/etc/docker/daemon.json（若该文件不存在则创建一个），修改内容如下：

```
[root@client ~]# vi /etc/docker/daemon.json
[root@client ~]# cat /etc/docker/daemon.json
{
"insecure-registries": ["192.168.200.11:5000"]
}
```

该配置文件内容的意思为指定 Registry 私有仓库服务器的 IP 地址及端口，修改完配置文件后，重启 Docker 服务，并查看私有仓库中的镜像及版本，最后拉取镜像，具体命令如下：

```
[root@client ~]# systemctl restart docker
[root@client ~]# curl 192.168.200.11:5000/v2/_catalog
{"repositories":["nginx"]}
[root@client ~]# curl 192.168.200.11:5000/v2/nginx/tags/list
{"name":"nginx","tags":["latest"]}
[root@client ~]# docker pull 192.168.200.11:5000/nginx:latest
latest: Pulling from nginx
d121f8d1c412: Pull complete
66a200539fd6: Pull complete
e9738820db15: Pull complete
d74ea5811e8a: Pull complete
ffdacbba6928: Pull complete
Digest: sha256:416d511ffa63777489af47f250b70d1570e428b67666567085f2bece3571ad83
Status: Downloaded newer image for 192.168.200.11:5000/nginx:latest
192.168.200.11:5000/nginx:latest
[root@client ~]# docker images
REPOSITORY                   TAG       IMAGE ID       CREATED        SIZE
192.168.200.11:5000/nginx    latest    992e3b7be046   3 months ago   133MB
```

最后验证 client 端拉取私有仓库镜像是否成功。

3. Harbor 镜像仓库的安装与使用

（1）Harbor 仓库安装

Harbor 的 3 种安装方式如下。

① 在线安装：从 Docker Hub 下载 Harbor 相关镜像，该安装软件包非常小。

② 离线安装：安装包包含部署的相关镜像，该安装包比较大。

③ OVA 安装程序（第三方）：当用户具有 vCenter 环境时，使用此安装程序，在部署 OVA 后启动 Harbor。

此处使用离线安装的方式，部署 Harbor 仓库服务。在安装 Harbor 之前，首先需要安装 docker-compose 服务，将提供的 docker-compose 目录上传到 master 节点的/root 目录下，然后将目录中的 v1.25.5-docker-compose-Linux-x86_64 文件放到/usr/local/bin/目录下并改名为 docker-compose，赋予执行权限，具体命令如下：

```
[root@master ~]# cp docker-compose/v1.25.5-docker-compose-Linux-x86_64 /usr/
```

```
local/bin/docker-compose
    [root@master ~]# chmod +x /usr/local/bin/docker-compose
    [root@master ~]# docker-compose --version
    docker-compose version 1.25.5, build 8a1c60f6
```

使用 docker-compose --version 命令验证 docker-compose 安装是否已完成。

将提供的 harbor-offline-installer-v2.1.0.tgz 软件包上传至 master 节点的/root 目录下，并解压，命令如下：

```
[root@master ~]# tar -zxf harbor-offline-installer-v2.1.0.tgz -C /usr/local/
[root@master ~]# cd /usr/local/harbor/
[root@master harbor]# ll
total 545264
-rw-r--r--. 1 root root       3361 Sep 15 22:48 common.sh
-rw-r--r--. 1 root root  558317240 Sep 15 22:49 harbor.v2.1.0.tar.gz
-rw-r--r--. 1 root root       8136 Sep 15 22:48 harbor.yml.tmpl
-rwxr-xr-x. 1 root root       2523 Sep 15 22:48 install.sh
-rw-r--r--. 1 root root      11347 Sep 15 22:48 LICENSE
-rwxr-xr-x. 1 root root       1881 Sep 15 22:48 prepare
```

Harbor-v2.1 版本跟以前不一样，没有了 harbor.cfg 文件，需要手动复制 harbor.yml.tmpl 再进行修改，命令如下：

```
[root@master harbor]# cp harbor.yml.tmpl harbor.yml
[root@master harbor]# vi harbor.yml
# Configuration file of Harbor

# The IP address or hostname to access admin UI and registry service.
# DO NOT use localhost or 127.0.0.1, because Harbor needs to be accessed by external clients.
hostname: 192.168.200.11                    //此处改成主机的 IP

# http related config
http:
  # port for http, default is 80. If https enabled, this port will redirect to https port
  port: 80

# https related config
#https:                                      //把关于 https 的参数全部注释
  # https port for harbor, default is 443
#  port: 443
  # The path of cert and key files for nginx
#  certificate: /your/certificate/path
#  private_key: /your/private/key/path
```

安装 Harbor 之前，首先需要将部署的 Registry 仓库删除，命令如下：

```
[root@master harbor]# docker ps -a
  CONTAINER ID        IMAGE                 COMMAND                  CREATED
STATUS                PORTS                 NAMES
  584a017941fe        registry:latest       "/entrypoint.sh /
etc…"    25 hours ago        Up About a minute       0.0.0.0:5000->5000/tcp
registry
[root@master harbor]# docker rm -f 584a017941fe
  584a017941fe
```

然后使用如下命令部署安装 Harbor：

```
[root@master harbor]# ./prepare
[root@master harbor]# ./install.sh --with-clair
... ...
[Step 5]: starting Harbor ...
Creating network "harbor_harbor" with the default driver
Creating network "harbor_harbor-clair" with the default driver
Creating harbor-log     ... done
Creating harbor-db      ... done
Creating harbor-portal  ... done
Creating registryctl    ... done
Creating registry       ... done
Creating redis          ... done
Creating clair          ... done
Creating harbor-core    ... done
Creating clair-adapter  ... done
Creating harbor-jobservice ... done
Creating nginx          ... done
✔ ----Harbor has been installed and started successfully.----
```

看到如上显示，即表明 Harbor 安装成功，使用浏览器访问 http://192.168.200.11，即能访问 Harbor 首页。登录 Harbor 仓库（用户名/密码：admin/Harbor12345），如图 2-3 和图 2-4 所示。

图 2-3　Harbor 登录界面

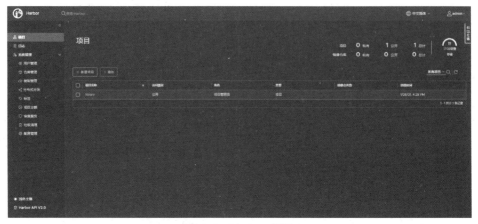

图 2-4　Harbor 首页

启动、停止和重启 Harbor 服务的命令如下：

```
# docker-compose up -d
# docker-compose stop
# docker-compose restart
```

（2）Harbor 镜像仓库上传测试

在上传镜像前，需要修改配置文件并重启 Docker 服务，命令如下：

```
[root@master ~]# vi /etc/docker/daemon.json
[root@master ~]# cat /etc/docker/daemon.json
{
"insecure-registries": ["192.168.200.11"]
}
```

修改原来的 daemon.json 文件，去掉 5000 端口，然后重启 Docker 服务，命令如下：

```
[root@master ~]# systemctl restart docker
```

重启完 Docker 服务之后，Harbor 服务会受到影响，需要到 Harbor 目录重新启动 Harbor 服务，命令如下：

```
[root@master harbor]# docker-compose start
Starting log           ... done
Starting registry      ... done
Starting registryctl   ... done
Starting postgresql    ... done
Starting portal        ... done
Starting redis         ... done
Starting core          ... done
Starting jobservice    ... done
Starting proxy         ... done
Starting clair         ... done
Starting clair-adapter ... done
```

上传至 Harbor 仓库的镜像与上传至 Registry 仓库的镜像在打标签（tag）的时候它们的方式是不同的，下面给 Nginx 镜像重新打标签，使其可以上传至 Harbor 仓库中，命令如下：

```
[root@master harbor]# docker tag 992e3b7be046 192.168.200.11/library/nginx:latest
```

这种打标签的方式取消了 5000 端口，增加了 library 目录，该目录为 Harbor 中的默认项目目录，若想推送到另外一个项目，可以在 Harbor 界面先创建项目，再打标签。

在推送镜像前，首先还需要登录 Harbor，具体命令如下：

```
[root@master harbor]# docker login -u admin -p Harbor12345 192.168.200.11
WARNING! Using --password via the CLI is insecure. Use --password-stdin.
WARNING! Your password will be stored unencrypted in /root/.docker/config.json.
Configure a credential helper to remove this warning. See
https://docs.docker.com/engine/reference/commandline/login/#credentials-store

Login Succeeded
[root@master harbor]# docker push 192.168.200.11/library/nginx:latest
The push refers to repository [192.168.200.11/library/nginx]
8032102adebe: Pushed
8eb80f066de2: Pushed
7230cfe05cc1: Pushed
822ae9fef1d8: Pushed
07cab4339852: Pushed
latest: digest: sha256:416d511ffa63777489af47f250b70d1570e428b67666567085f2bece3571ad83 size: 1362
```

可以在 Harbor 界面中查看镜像是否上传成功，如图 2-5 所示。

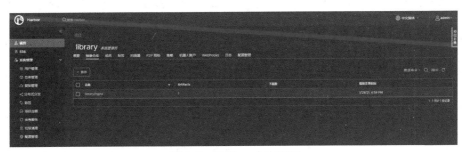

图 2-5　Harbor 仓库界面

至此，测试上传镜像至 Harbor 仓库成功。

（3）client 节点测试拉取镜像

首先修改 Docker 的配置文件/etc/docker/daemon.json，并重启 Docker 服务，命令如下：

```
[root@client ~]# vi /etc/docker/daemon.json
[root@client ~]# cat /etc/docker/daemon.json
```

```
{
"insecure-registries": ["192.168.200.11"]
}
[root@client ~]# systemctl restart docker
```

若想在 Harbor 仓库的 client 端 pull 镜像，也需要进行登录操作，命令如下：

```
[root@client ~]# docker login -u admin -p Harbor12345 192.168.200.11
WARNING! Using --password via the CLI is insecure. Use --password-stdin.
WARNING! Your password will be stored unencrypted in /root/.docker/config.json.
Configure a credential helper to remove this warning. See
https://docs.docker.com/engine/reference/commandline/login/#credentials-store

Login Succeeded
```

登录成功后，使用 docker pull 命令下载镜像，目前仓库中只有 Nginx 镜像，进行下载测试，命令如下（在下载镜像前，可以先将 client 节点现有的镜像删除）：

```
[root@client ~]# docker pull 192.168.200.11/library/nginx
Using default tag: latest
latest: Pulling from library/nginx
d121f8d1c412: Pull complete
66a200539fd6: Pull complete
e9738820db15: Pull complete
d74ea5811e8a: Pull complete
ffdacbba6928: Pull complete
Digest: sha256:416d511ffa63777489af47f250b70d1570e428b67666567085f2bece3571ad83
Status: Downloaded newer image for 192.168.200.11/library/nginx:latest
192.168.200.11/library/nginx:latest
[root@client ~]# docker images
REPOSITORY                           TAG      IMAGE ID       CREATED        SIZE
192.168.200.11/library/nginx         latest   992e3b7be046   3 months ago   133MB
```

client 端下载镜像测试成功。关于 Harbor 仓库的更多功能，感兴趣的同学可以自行学习。

归纳总结

通过本单元内容的学习，掌握了 Registry 和 Harbor 镜像仓库的应用场景与特点。还通过实操练习，掌握了两种镜像仓库的部署方法与使用命令，读者可以通过镜像包部署属于自己的镜像仓库并使用。在日常工作中，部署属于自己的镜像仓库是非常普遍的，拥有自己的镜像仓库，会让工作事半功倍。

课后练习

一、判断题

1. 进程隔离是 Cgroup 的功能。（   ）
2. Docker 拉取镜像的命令是 docker pull。（   ）

二、单项选择题

1. 在 Docker 镜像仓库中搜索镜像会用到以下（   ）命令。

A. docker search　　　B. docker push　　　C. docker ps　　　D. docker pull

2. Harbor 私有镜像仓库是以下（   ）公司的项目。

A. Google　　　B. Redhat　　　C. Microsoft　　　D. VMware

三、多项选择题

1. Docker 架构包含了以下（   ）部分。

A. Containers　　　B. Images　　　C. Docker Daemon　　　D. REST API

2. 下面属于 Docker Swarm 的特点的有（   ）。

A. Docker Engine 集成集群管理　　　B. 扩容缩容

C. 状态动态协调及负载均衡　　　D. 多主机网络及服务发现

技能训练

1. 安装 Registry 仓库容器服务，并上传镜像。
2. 安装部署 Docker 引擎和 Docker 仓库，采用单节点方式，使用本地源安装 docker-ce，配置好私有镜像仓库 Harbor。

单元 3　持久化存储卷

学习目标

通过本单元的学习，要求了解容器的持久化存储卷的概念、存储卷的作用、使用存储卷的优势与优点、不同方式的持久化存储卷实现方法、存储卷的使用场景与使用方法等。本单元培养读者掌握 Bind mounts 持久化存储卷的实现与使用、Volume 持久化存储卷的实现与使用等技能，也培养读者的自我学习能力和逻辑思维能力。

3.1　认识容器存储卷

1. 持久化存储卷介绍

Docker 的镜像是只读的，但是容器是可写的，可以将数据写入容器，不过一旦容器被删除，数据将会丢失。那么有什么办法能将数据进行持久化存储呢？人们都知道 Docker 的数据可以存在容器的可写层，但是当容器不再运行时，数据就不能持久存储，无法从容器外部获得这些数据；并且，Docker 的可写层与宿主机（Host）在容器运行时紧密耦合，不能轻易地把数据移动到别的地方。

Docker 提供了 3 种方式把数据从 Host 挂载到 Docker 容器中，并实现数据存储。这 3 种方式分别是 Bind mounts、Volume、Tmpfs。其中 Bind mounts、Volume 可以实现持久化存储，而 Tmpfs 不能实现持久化存储，如图 3-1 所示。

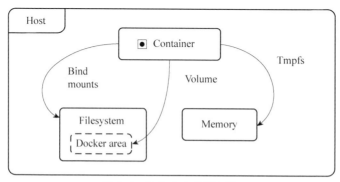

图 3-1　持久化存储卷架构

Bind mounts：可以实现将 Host 中的任意路径挂载到容器中，也可以存储在任意位置，但非 Docker 的进程或者容器可能随时对其进行修改，存在一定的风险。

Volume：可以实现将 Docker 中的某一个 Volume 挂载到容器中，存储在 Docker Host 文件系统的某一个路径下，默认是/var/lib/docker/volumes，由 Docker 进行管理，非 Docker 的进程不能修改该路径下的文件，比较安全。

Tmpfs：存储在 Host 系统的内存中，没有在磁盘上进行存储，不能实现持久化。

2. Bind mounts 实现持久化存储

Bind mounts 可以将 Host 上的任意路径进行挂载，包括重要的文件系统路径，因此这就存在一个缺点，即用户可以通过容器更改 Host 的文件系统，进行文件的增删改等操作，若将文件系统的路径进行挂载，则会严重影响到系统的安全性，包括那些非 Docker 进程。

使用 Bind mounts 方式做数据卷的映射时，首次使用 docker run –v 命令运行，如果宿主机的文件夹没有内容，Docker 容器中的文件夹有内容，则宿主机的内容会覆盖 Dokcer 容器中的内容，也就是容器中原本有内容的也会变为没有内容。

Bind mounts 只能通过 docker run –v 命令方式启动，无法使用 Dockerfile 文件的方式。运行容器的时候会指定本地的一个文件目录和容器中的一个文件目录进行映射，通过这个可以使文件数据同步，双方无论哪一方有修改，另一方都会同步更新。

在单容器情况下，使用 Bind mounts 时可以添加 "-v" 或者 "--volume" 选项，在 Docker17.06 之后，可以统一使用 "--mount"，这也是推荐使用的。

-v 或者--volume 语法如下：它由三部分组成，使用 ":" 进行分割，这些字段必须以正确的顺序排列，并且每个字段的含义不明显。

第一个字段是 Docker Host（宿主机）上的一个文件或者目录。

第二个字段是将要挂载到容器上的一个文件或者目录。

第三个字段是可选的，用来增加一些附加选项，比如 ro、consistent、delegated、cached、z 或 Z 等。

--mount 语法如下：它由一组键值对组成，由 "," 进行分割，mount 语法如表 3-1 所示。

表 3-1 mount 语法

Key	Value
type	Bind mounts、Volume、Tmpfs，如不指定，默认是 Volume
source 或者 src	Docker Host 上的一个文件或者目录
destination 或者 dst 或者 target	被挂载容器上的一个文件或者目录
readonly	默认为可读可写，加上 readonly 表示只读的文件系统
bind-propagation	rprivate、private、rshared、shared、rslave、slave

-v 和--volume 两者的区别如下：

使用-v 的时候，如果在 Docker Host 中不存在要挂载的文件或者目录，Docker 将会自动创建，通常是一个目录。

使用--mount 的时候，如果在 Docker Host 中不存在要挂载的文件或者目录，Docker 不会自动创建目录，并生成一个错误。

使用场景：

① 把 Host 中的配置文件共享给 Host 上面的容器。容器为什么自带 DNS 解析呢？那

是因为默认情况下 Host 把/etc/resolv.conf 挂载到它上面的容器里面了。

② 在 Docker Host 上面的开发环境和容器直接共享程序的源代码或者构建要素。例如，用户可以挂载一个 Maven 目录到一个容器中，每当用户在 Docker Host 重新建立 Maven 项目，容器都可以直接获取用户重新构建的 Maven 项目。

③ 可以将源代码目录挂载到容器中，在 Host 中修改代码就能看到应用的实时效果。

④ 将 MySQL 容器的数据放在 Bind mounts 里，这样 Host 可以方便地备份和迁移数据。

⑤ 只需要向容器添加文件，不需要覆盖整个目录。

3. Volume 实现持久化存储

Volume 完全由 Docker 进行管理，比如 Volume 的创建，用户可以使用命令 docker volume create 来简单地创建一个 Volume，当容器或者服务被创建的时候，Docker 也可以自动地创建一个 Volume。

用户创建的 Volume 将存储在 Docker Host（宿主机）的存储目录下，并由 Docker 进行管理。

一个给定的 Volume 可以同时挂载到多个容器中。当没有容器使用 Volume 时，Volume 对 Docker 仍然是可用的，并且不会被自动删除，使用 docker volume rm 命令可以删除 Volume。

在挂载 Volume 时，可以对其命名，也可以默认其随机生成的名字。如果用户没有指定名称，则当 Volume 第一次挂载到一个容器时，Docker 会用一个随机字符串对其进行命名，这样可以保证 Volume 在 Docker Host（宿主机）中的唯一性。

Volume 还支持使用 Volume Drivers。

4. Volume 对比 Bind mounts 方式的优点

Volume 的备份和迁移更加容易。

可以使用 Docker CLI 或者 Docker API 管理 Volume。

Volume 既可以在 Linux 的容器中使用，也可以在 Windows 的容器中使用。

Volume 在多容器中共享更加安全。

Volume Drivers 允许用户把数据存储在远程主机或云提供商上。

5. Volume 的使用场景

当用户需要在多个正在运行的容器之间共享数据时，需要使用 Volume。如果用户没有明确指定创建它，那么它在第一次装入容器时就会创建一个 Volume。当容器停止或被删除时，Volume 仍然存在。多个容器可以同时读写一个 Volume。只有当用户明确指定要删除某个 Volume 时，它才会被删除。

当用户需要把容器的数据永久存储在一个远程主机或者一个云服务器上时，需要使用 Volume。

当用户的 Docker Host 无法保证可以提供一个目录或者文件来存储数据时，也需要 Volume，它可以减少用户对配置文件的依赖。

当用户需要备份数据，或者恢复数据，以及需要把数据从一个 Docker Host 迁移到另外一个 Docker Host 的时候，Volume 是最好的选择。用户可以停掉正在使用 Volume 的容器，然后把 Volume 的目录备份下来即可，Volume 的目录一般在/var/lib/docker/volumes/下。

3.2 实现存储卷持久化

1. 环境准备

（1）规划节点

存储卷节点的规划如表 3-2 所示。

表 3-2　存储卷节点的规划

IP 地址	主机名	节点
192.168.200.13	docker	Docker 实验节点

（2）基础准备

准备好 VMware Workstation 工具和 CentOS7.5 虚拟机快照，克隆一台虚拟机，云主机配置自定义，推荐配置 2vCPU、4GB 内存、40GB 硬盘，网络使用 NAT 模式，Docker 节点 IP 设置为 192.168.200.13，推荐设置主机密码为 000000，使用克隆的虚拟机作为实验基础环境。注意：与之前类似，在进行持久化存储卷实验前，自行将虚拟机配置好 IP、安装 Docker 服务，关闭 SELinux 和防火墙服务（安装 Docker 服务的实验步骤在 1.2 节中）。

2. Bind mounts 持久化存储卷使用

（1）Bind mounts 正常使用

将提供的 nginx_latest.tar 镜像包，上传至服务器的/root 目录下，并使用 load 命令加载，具体命令如下：

```
[root@docker ~]# docker load -i nginx_latest.tar
07cab4339852: Loading layer [==================>]   72.49MB/72.49MB
822ae9fef1d8: Loading layer [==================>]   64.53MB/64.53MB
7230cfe05cc1: Loading layer [==================>]   3.072kB/3.072kB
8eb80f066de2: Loading layer [==================>]   4.096kB/4.096kB
8032102adebe: Loading layer [==================>]   3.584kB/3.584kB
Loaded image: nginx:latest
```

使用命令查看镜像列表，命令如下：

```
[root@docker ~]# docker images
REPOSITORY      TAG         IMAGE ID        CREATED         SIZE
nginx           latest      992e3b7be046    4 months ago    133MB
```

在 Host 上创建目录（因为通过--mount 进行挂载时要求用户在 Host 上有这个目录），命令如下：

```
[root@docker ~]# mkdir -p /root/nginx-test/html
```

运行容器，将宿主机的目录/root/nginx-test/html 与 Nginx 容器内部工作目录/usr/share/nginx/html 相连，具体命令如下：

```
[root@docker ~]# docker run -itd --name nginxtest -p 80:80 --mount type=bind,
source=/root/nginx-test/html,target=/usr/share/nginx/html nginx:latest
    7e74de9533cb3ea4b4e6b98ce1e18fd65734572fece7d7d00ab4b61f28be0b33
```

使用 docker ps -a 命令查看容器运行情况，命令如下：

```
[root@docker ~]# docker ps -a
CONTAINER ID        IMAGE              COMMAND                  CREATED
STATUS              PORTS              NAMES
7e74de9533cb        nginx:latest       "/docker-entrypoint.…"   29 seconds
ago     Up 28 seconds      0.0.0.0:80->80/tcp    nginxtest
```

使用 docker inspect 命令查看容器是否正确挂载，Source 为宿主机上的路径，Destination 为容器上的路径，RW 表示可读可写的模式，也是默认的模式，命令如下（inspect 命令显示的内容较多，此处只截取 Mounts 字段的内容展示）：

```
[root@docker ~]# docker inspect nginxtest
"Mounts": [
    {
        "Type": "bind",
        "Source": "/root/nginx-test/html",
        "Destination": "/usr/share/nginx/html",
        "Mode": "",
        "RW": true,
        "Propagation": "rprivate"
    }
```

可以进入容器内部查看 Nginx 的项目目录为空（因为 Host 中相应的路径下没有内容），命令如下：

```
[root@docker ~]# docker exec -it nginxtest /bin/bash
root@7e74de9533cb:/# ls /usr/share/nginx/html/
root@7e74de9533cb:/#
```

此时宿主机的/root/nginx-test/html 目录与容器内的/usr/share/nginx/html 目录是相连的，在宿主机的/root/nginx-test/html 目录内添加或修改的内容，会同步到容器内的/usr/share/nginx/html 目录。在宿主机/root/nginx-test/html 目录下创建 index.html 文件，并编辑该文件，具体命令如下：

```
[root@docker ~]# vi nginx-test/html/index.html
[root@docker ~]# cat nginx-test/html/index.html
This is a bind mounts test !!!
```

可以通过页面访问进入容器内部验证，此处使用 curl 命令查看 Nginx 首页内容进行验证，具体命令如下：

```
[root@docker ~]# curl 127.0.0.1
This is a bind mounts test !!!
```

可以通过停止容器、删除容器验证文件是否还存在，命令如下：

```
[root@docker ~]# docker stop 7e74de9533cb
7e74de9533cb
[root@docker ~]# curl 127.0.0.1
curl: (7) Failed connect to 127.0.0.1:80; Connection refused
[root@docker ~]# docker rm 7e74de9533cb
7e74de9533cb
[root@docker ~]# ll nginx-test/html/
total 4
-rw-r--r--. 1 root root 31 Feb 25 21:08 index.html
[root@docker ~]# cat nginx-test/html/index.html
This is a bind mounts test !!!
```

从上面的操作命令及返回结果可以看出，关闭容器后，便不能访问 80 端口了；然而在删除容器之后，index.html 文件还在宿主机中，因此该方法（Bind mounts）可以实现存储持久化。

（2）Bind mounts 只读模式

若用户仅仅是想读取 Host 宿主机上的文件，而不想修改其中的内容，则可以将默认的读写模式修改为只读模式，命令如下：

```
[root@docker ~]# docker run -itd --name nginxro -p 80:80 --mount type=bind,source=/root/nginx-test/html,target=/usr/share/nginx/html,readonly nginx:latest
c062f59dbc44409d6ef1651063623ae8397675dce7c9b575dff32617bc780031
```

启动之后，可以通过 docker inspect 命令查看容器中的 Mounts 信息，具体命令如下：

```
[root@docker ~]# docker inspect nginxro
"Mounts": [
        {
            "Type": "bind",
            "Source": "/root/nginx-test/html",
            "Destination": "/usr/share/nginx/html",
            "Mode": "",
            "RW": false,
            "Propagation": "rprivate"
        }
```

可以看到此时的"RW"字段为 false。

进入容器内部，对 Nginx 项目目录/usr/share/nginx/html 下的 index.html 进行修改，会发现不允许修改文件（同样的操作在默认使用 bind mounts 的时候不会发生），具体命令如下：

```
[root@docker ~]# docker exec -it c062f59dbc44 /bin/bash
root@c062f59dbc44:/# cd /usr/share/nginx/html/
root@c062f59dbc44:/usr/share/nginx/html# ls
index.html
root@c062f59dbc44:/usr/share/nginx/html# cat index.html
This is a bind mounts test !!!
root@c062f59dbc44:/usr/share/nginx/html# echo "write test" >> index.html
bash: index.html: Read-only file system
root@c062f59dbc44:/usr/share/nginx/html#
```

回到宿主机，修改 index.html 文件并查看，命令如下：

```
[root@docker ~]# cd nginx-test/html/
[root@docker html]# echo "write test" >> index.html
[root@docker html]# curl 127.0.0.1
This is a bind mounts test !!!
write test
```

然后再进入容器内部查看，命令如下：

```
[root@docker html]# docker exec -it c062f59dbc44 /bin/bash
root@c062f59dbc44:/# cat /usr/share/nginx/html/index.html
This is a bind mounts test !!!
write test
```

使用 Bind mounts 只读模式时宿主机可以对文件进行读写，而容器内部只拥有只读权限，以上就是 Bind mounts 使用只读模式实现存储持久化的案例。

3. Volume 持久化存储卷使用

（1）Volume 使用方式一

使用 MySQL 镜像进行实验，MySQL 镜像中使用了容器的 Volume 持久化存储（可以通过官方 MySQL 镜像的 Dockerfile 查看，当中用到了 Volume 持久化存储技术，查看地址为 https://github.com/docker-library/mysql/blob/master/8.0/Dockerfile.debian），Dockerfile 内容部分截取如下：

```
# ensure that /var/run/mysqld (used for socket and lock files) is writable
regardless of the UID our mysqld instance ends up having at runtime
  && chmod 1777 /var/run/mysqld /var/lib/mysql

VOLUME /var/lib/mysql


# Config files
COPY config/ /etc/mysql/
COPY docker-entrypoint.sh /usr/local/bin/
RUN ln -s usr/local/bin/docker-entrypoint.sh /entrypoint.sh # backwards
compat
```

```
ENTRYPOINT ["docker-entrypoint.sh"]

EXPOSE 3306 33060
CMD ["mysqld"]
```

使用提供的 mysql-latest.tar 上传至宿主机的/root 目录下,并将该镜像加载,命令如下:

```
[root@docker ~]# docker load -i mysql-latest.tar
9eb82f04c782: Loading layer [==================>]  72.49MB/72.49MB
6427e36e3873: Loading layer [==================>]  338.4kB/338.4kB
1a4bac3ca47c: Loading layer [==================>]  9.539MB/9.539MB
7c06655789f2: Loading layer [==================>]  4.201MB/4.201MB
411bb8845b78: Loading layer [==================>]  1.536kB/1.536kB
902a5f810181: Loading layer [==================>]  53.76MB/53.76MB
52d9c23a9a20: Loading layer [==================>]  6.656kB/6.656kB
9ffac285cc4f: Loading layer [==================>]  3.584kB/3.584kB
5ba1eb4d2ba5: Loading layer [==================>]  411MB/411MB
9ba725af2ff0: Loading layer [==================>]  5.632kB/5.632kB
7cdc784cbcfc: Loading layer [==================>]  16.38kB/16.38kB
9fda8797bc31: Loading layer [==================>]  1.536kB/1.536kB
Loaded image: mysql:latest
[root@docker ~]# docker images
REPOSITORY      TAG        IMAGE ID         CREATED        SIZE
mysql           latest     2933adc350f3     2 weeks ago    546MB
nginx           latest     992e3b7be046     4 months ago   133MB
```

使用命令查看当前环境下的容器 Volume,具体命令如下:

```
[root@docker ~]# docker volume ls
DRIVER              VOLUME NAME
```

可以看到当前环境下没有 Volume,使用 MySQL 镜像启动容器,再次查看 Volume,命令如下:

```
[root@docker ~]# docker run -d --name mysql1 -e MYSQL_ALLOW_EMPTY_PASSWORD=true mysql:latest
fee3c4f9ffd12afc379c7053036253375e0f76bf3f094dc4a3c4a7b31022886e
[root@docker ~]# docker ps -a
CONTAINER ID        IMAGE               COMMAND                  CREATED             STATUS              PORTS                     NAMES
fee3c4f9ffd1        mysql:latest        "docker-entrypoint.s…"   3 seconds ago       Up 2 seconds        3306/tcp, 33060/tcp       mysql1
[root@docker ~]# docker volume ls
DRIVER     VOLUME NAME
local      d5e2619dc1ef20828adaafdc22e75650fcbcf6df221787f8a8c79bc3d5aabad5
```

查看该容器 Volume 的详细信息,命令如下:

```
[root@docker ~]# docker volume inspect d5e2619dc1ef20828adaafdc22e75650fcbc
f6df221787f8a8c79bc3d5aabad5
[
    {
        "CreatedAt": "2021-03-01T01:22:54-05:00",
        "Driver": "local",
        "Labels": null,
        "Mountpoint": "/var/lib/docker/volumes/d5e2619dc1ef20828adaafdc22e75650fcbcf6df221787f8a8c79bc3d5aabad5/_data",
        "Name": "d5e2619dc1ef20828adaafdc22e75650fcbcf6df221787f8a8c79bc3d5aabad5",
        "Options": null,
        "Scope": "local"
    }
]
```

可以看到，这个 Volume 并不在容器当中，而是挂载到了宿主机的/var/lib/docker/volumes/d5e2619dc1ef20828adaafdc22e75650fcbcf6df221787f8a8c79bc3d5aabad5/_data 目录下。

删除这个容器，命令如下：

```
[root@docker ~]# docker stop mysql1
mysql1
[root@docker ~]# docker rm mysql1
mysql1
```

删除容器后，查看容器 Volume 是否还存在，命令如下：

```
[root@docker ~]# docker volume ls
DRIVER      VOLUME NAME
local       d5e2619dc1ef20828adaafdc22e75650fcbcf6df221787f8a8c79bc3d5aabad5
[root@docker ~]# ll /var/lib/docker/volumes/
total 24
drwxr-xr-x. 3 root root     19 Mar  1 01:22 d5e2619dc1ef20828adaafdc22e75650fc
bcf6df221787f8a8c79bc3d5aabad5
-rw-------. 1 root root  32768 Mar  1 01:22 metadata.db
```

可以发现，Volume 在删除容器之后也不会被删除，可以防止删除容器之后数据也不存在了的问题。但是 Volume 的名字非常不友好，不方便使用，用户可以在创建或者启动容器的时候给 Volume 设置别名来方便使用。具体命令如下：

```
[root@docker ~]# docker run -d -v mysql-vol:/var/lib/mysql --name mysql1 -e MYSQL_ALLOW_EMPTY_PASSWORD=true mysql:latest
afb6669a83c660baf983397170c8211b127e88ab26724d6217f9f8f567bc998f
[root@docker ~]# docker volume ls
DRIVER      VOLUME NAME
local       d5e2619dc1ef20828adaafdc22e75650fcbcf6df221787f8a8c79bc3d5aabad5
```

```
local         mysql-vol
[root@docker ~]# ll /var/lib/docker/volumes/
total 24
-rw-------. 1 root root 32768 Mar  1 04:07 metadata.db
drwxr-xr-x. 3 root root    19 Mar  1 04:07 mysql-vol
```

可以看到，Volume 的名字变成了修改后的 mysql-vol，这个名字方便管理与使用。使用 mysql-vol 这个 Volume 验证持久化存储，进入数据库容器，并创建数据库，命令如下：

```
[root@docker ~]# docker exec -it mysql1 /bin/bash
root@afb6669a83c6:/# mysql
Welcome to the MySQL monitor.  Commands end with ; or \g.
Your MySQL connection id is 8
Server version: 8.0.23 MySQL Community Server - GPL

Copyright (c) 2000, 2021, Oracle and/or its affiliates.

Oracle is a registered trademark of Oracle Corporation and/or its
affiliates. Other names may be trademarks of their respective
owners.

Type 'help;' or '\h' for help. Type '\c' to clear the current input statement.

mysql> show databases;
+--------------------+
| Database           |
+--------------------+
| information_schema |
| mysql              |
| performance_schema |
| sys                |
+--------------------+
4 rows in set (0.01 sec)
mysql> create database docker;
Query OK, 1 row affected (0.00 sec)

mysql> show databases;
+--------------------+
| Database           |
+--------------------+
| docker             |
| information_schema |
| mysql              |
| performance_schema |
| sys                |
```

```
+--------------------+
5 rows in set (0.00 sec)
```

停止 mysql1 容器,并删除 mysql1 容器,命令如下:

```
[root@docker ~]# docker stop mysql1
mysql1
[root@docker ~]# docker rm mysql1
mysql1
```

重新创建一个 mysql2 容器,继续使用 mysql-vol 这个 Volume,验证之前的数据是否还存在,命令如下:

```
[root@docker ~]# docker run -d -v mysql-vol:/var/lib/mysql --name mysql2 -e MYSQL_ALLOW_EMPTY_PASSWORD=true mysql:latest
bd729db6c77e9c39f4544963c5c896ac236f2d209d00b9c43cbba71479f813bc
[root@docker ~]# docker volume ls
DRIVER      VOLUME NAME
local       d5e2619dc1ef20828adaafdc22e75650fcbcf6df221787f8a8c79bc3d5aabad5
local       mysql-vol
```

可以看到,创建容器后,并没有新的 Volume 被创建。进入容器,查看数据库中的内容,命令如下:

```
[root@docker ~]# docker exec -it mysql2 /bin/bash
root@bd729db6c77e:/# mysql
Welcome to the MySQL monitor.  Commands end with ; or \g.
Your MySQL connection id is 8
Server version: 8.0.23 MySQL Community Server - GPL

Copyright (c) 2000, 2021, Oracle and/or its affiliates.

Oracle is a registered trademark of Oracle Corporation and/or its
affiliates. Other names may be trademarks of their respective
owners.

Type 'help;' or '\h' for help. Type '\c' to clear the current input statement.

mysql> show databases;
+--------------------+
| Database           |
+--------------------+
| docker             |
| information_schema |
| mysql              |
| performance_schema |
```

```
| sys                |
+--------------------+
5 rows in set (0.01 sec)
```

可以看到数据库中存在 Docker 数据库，Volume 中的数据并不会因为容器的删除而消失，从而实现了数据持久化的目标。但是这种方式的 Volume 需要在 Dockerfile 中使用 Volume 来预先指定容器中的数据存放路径。下面使用自行创建 Volume 的方式来实现持久化存储。

（2）Volume 使用方式二

在使用自行创建的 Volume 之前，先删除当前存在的容器与 Volume，确保环境是干净的，命令不再赘述。删除后，使用命令查看当前环境，命令如下：

```
[root@docker ~]# docker ps -a
CONTAINER ID        IMAGE                   COMMAND                 CREATED
STATUS              PORTS              NAMES
[root@docker ~]# docker volume ls
DRIVER              VOLUME NAME
```

创建一个 Volume，并查看 Volume 的详细信息，命令如下：

```
[root@docker ~]# docker volume create test_vol
test_vol
[root@docker ~]# docker volume inspect test_vol
[
    {
        "CreatedAt": "2021-03-01T22:08:44-05:00",
        "Driver": "local",
        "Labels": {},
        "Mountpoint": "/var/lib/docker/volumes/test_vol/_data",
        "Name": "test_vol",
        "Options": {},
        "Scope": "local"
    }
]
```

刚刚创建了一个名为 test_vol 的 Volume，进入以下的路径查看，可以看到这是一个空的卷，命令如下：

```
[root@docker ~]# ll /var/lib/docker/volumes/test_vol/_data/
total 0
```

使用这个空的 Volume 启动一个容器，将提供的 Nginx 容器上传至服务器内并 load（命令不再赘述），启动 Nginx 容器并使用 test_vol 这个 Volume，命令如下：

```
[root@docker ~]# docker run -itd --name nginxtest -p 80:80 --mount
source=test_vol,target=/usr/share/nginx/html nginx:latest
    f32eafb58d9f934e93cf4718828666ded1c135f147495ee33173e75ae7e62488
[root@docker ~]# docker ps -a
```

```
    CONTAINER ID          IMAGE                COMMAND                  CREATED
STATUS              PORTS                NAMES
    f32eafb58d9f          nginx:latest         "/docker-entrypoint.…"   3 seconds
ago        Up 2 seconds         0.0.0.0:80->80/tcp   nginxtest
```

通过 docker inspect nginxtest 命令查看容器的详细情况，命令如下（只截取"Mounts"部分）：

```
"Mounts": [
    {
        "Type": "volume",
        "Name": "test_vol",
        "Source": "/var/lib/docker/volumes/test_vol/_data",
        "Destination": "/usr/share/nginx/html",
        "Driver": "local",
        "Mode": "z",
        "RW": true,
        "Propagation": ""
    }
],
```

再次查看宿主机上 test_vol 卷中的内容，命令如下：

```
[root@docker ~]# ll /var/lib/docker/volumes/test_vol/_data/
total 8
-rw-r--r--. 1 root root 494 Sep 29 10:12 50x.html
-rw-r--r--. 1 root root 612 Sep 29 10:12 index.html
```

发现这里多了两个文件，因为用户用一个空的 Volume 来启动容器时，会将容器中的内容复制到用户的宿主机上。

删除容器，并查看 Volume 是否还存在，命令如下：

```
[root@docker ~]# docker rm -f nginxtest
nginxtest
[root@docker ~]# docker volume ls
DRIVER              VOLUME NAME
local               test_vol
```

可以看到 Volume 依然存在，Volume 中的内容也依旧存在，命令如下：

```
[root@docker ~]# ll /var/lib/docker/volumes/test_vol/_data/
total 8
-rw-r--r--. 1 root root 494 Sep 29 10:12 50x.html
-rw-r--r--. 1 root root 612 Sep 29 10:12 index.html
```

接下来使用有数据的 Volume 来启动容器，首先将 test_vol 中的文件进行修改，删除 50x.html，然后编辑 index.html，具体命令如下：

```
[root@docker ~]# cd /var/lib/docker/volumes/test_vol/_data/
[root@docker _data]# ll
total 8
-rw-r--r--. 1 root root 494 Sep 29 10:12 50x.html
-rw-r--r--. 1 root root 612 Sep 29 10:12 index.html
[root@docker _data]# rm -rf 50x.html
[root@docker _data]# echo "this is a volume test" > index.html
[root@docker _data]# cat index.html
this is a volume test
```

使用 test_vol 这个 Volume 启动 Nginx 容器，并查看效果，命令如下：

```
[root@docker _data]# docker run -itd --name nginxtest2 -p 80:80 --mount source=test_vol,target=/usr/share/nginx/html nginx:latest
9f0ed866b48caf4ef3c2fa957ce8149f5fcf3c23113998b8be0c4a89a60b06d8
[root@docker _data]# curl 127.0.0.1
this is a volume test
```

可以看到，用一个现有的 Volume 也是可以进行挂载的，并且会将 Volume 中的文件挂载到容器中，覆盖掉容器中的文件。

最后，尝试如下情况：如果在运行容器时不提前创建 Volume，将会是什么情况？运行 Nginx 容器，命令如下（source=test_vol2 是一个不存在的 Volume）：

```
[root@docker ~]# docker run -itd --name nginxtest3 -p 81:80 --mount source=test_vol2,target=/usr/share/nginx/html nginx:latest
462c36761bee846ed44c750d47ef00adce045cb4110e25676d39bde7c0a5fdfc
```

启动容器后，查看 Volume 列表，命令如下：

```
[root@docker ~]# docker volume ls
DRIVER              VOLUME NAME
local               test_vol
local               test_vol2
[root@docker ~]# docker inspect test_vol2
[
    {
        "CreatedAt": "2021-03-02T00:34:56-05:00",
        "Driver": "local",
        "Labels": null,
        "Mountpoint": "/var/lib/docker/volumes/test_vol2/_data",
        "Name": "test_vol2",
        "Options": null,
        "Scope": "local"
    }
]
```

可以看到自动创建了一个名为 test_vol2 的 Volume，并且其存放位置也是 /var/lib/docker/volumes，因此也是受到 Docker 管理的。查看该 Volume 中的数据情况，能看到 Nginx 容器中的文件也直接复制到了宿主机 test_vol2 卷上，命令如下：

```
[root@docker ~]# ll /var/lib/docker/volumes/test_vol2/_data/
total 8
-rw-r--r--. 1 root root 494 Sep 29 10:12 50x.html
-rw-r--r--. 1 root root 612 Sep 29 10:12 index.html
```

下面对 Bind mounts 和 Volume 两种方式做一个对比，如表 3-3 所示。

表 3-3　Bind mounts 与 Volume 对比

不同点	Bind mounts	Volume
Source 位置	可以任意指定	/var/lib/docker/volumes/...
Host 源地址为空	覆盖掉容器的内容	容器内数据复制到 Volume
是否支持单个文件	支持	不支持，只能是目录
权限控制	读写或者只读	读写或者只读
移植性	弱，与 Host Path 绑定	强，无须指定 Host 目录

归纳总结

通过本单元内容的学习，掌握了持久化存储卷的使用场景与使用方法，读者可以使用 Bind mounts 或者 Volume 的方式实现持久化存储。在本单元中，还提到了 Tmpfs 方式的存储卷，感兴趣的读者可以自行寻找资料学习。

课后练习

一、判断题

1. Docker 创建容器的命令是 docker start。（　　）

2. Docker 停止容器的命令是 docker stop。（　　）

二、单项选择题

1. Docker 运行容器的命令是（　　）。

A. docker run　　　　B. docker-exec　　　　C. docker start　　　　D. docker attach

2. Docker 进入容器内部的命令是（　　）。

A. docker run　　　　B. docker-exec　　　　C. docker start　　　　D. docker attach

三、多项选择题

1. 下面关于 Docker Image 的说法中，正确的是（　　）。

A. 镜像是一个只读模板

B. 由 Dockerfile 文本描述镜像的内容

C. 镜像定义类似面向过程的类

D. 构建一个镜像实际上就是安装、配置和运行的过程
2. 下面关于 Docker 容器的说法中，正确的是（　　）。
A. 容器是一个镜像的运行实例
B. 可以通过运行用户指定的指令进行启动、停止、删除
C. 通过命令分配一个伪终端可以进入容器操作
D. 容器都是相互可见的

技能训练

1. 将宿主机目录成功挂载至容器目录。
2. 通过创建容器存储卷的方式，将存储卷挂载至容器 MySQL 内。

单元 4　容器网络

学习目标

通过本单元的学习，要求了解容器网络的概念、容器网络模式的几种原型，了解容器几种网络模式的特点、不同网络模式的特点与使用场景等。本单元培养读者掌握 Bridge、Host、None、Container 网络模式的实现方法等技能，也培养读者的创新意识和执行能力。

4.1 容器网络的几种原型

4.1.1 容器网络概述

随着容器知识的学习与使用，发现需要了解很多关于网络的知识，因此，用户有必要深入了解 Docker 的网络知识，以满足更高的网络需求。

首先简单描述一下容器的几种网络模式，容器共提供了 4 种网络模式，分别如下。

① Bridge 模式：此模式会为每一个容器分配、设置 IP 等，并将容器连接到一个 docker0 虚拟网桥，通过 docker0 网桥以及 iptables nat 表配置与宿主机通信。

② None 模式：该模式关闭了容器的网络功能。

③ Host 模式：容器将不会虚拟出自己的网卡、配置自己的 IP 等，而是使用宿主机的 IP 和端口。

④ Container 模式：创建的容器不会创建自己的网卡、配置自己的 IP，而是和一个指定的容器共享 IP、端口范围。

关于 4 种网络模式的对比，如表 4-1 所示。

表 4-1　容器的 4 种网络模式

Docker 网络模式	配置	说明	是否支持多主机
Bridge 模式	-net=bridge	默认模式	否
None 模式	-net=none	容器有独立的 Network Namespace，但并没有对其进行任何网络设置，如分配 veth pair 和网桥连接、配置 IP 等	否
Host 模式	-net=host	容器和宿主机共享 Network Namespace	是
Container 模式	-net=container:NAME_or_ID	容器和另外一个容器共享 Network Namespace。如 Kubernetes 中的 Pod 就是多个容器共享一个 Network Namespace	否

当安装好 Docker 引擎时默认会自动创建 3 个网络：Bridge、Host、None。可以使用一台已经安装好 Docker 环境的虚拟机，使用命令查看，如下所示：

```
[root@dockernet ~]# docker network list
NETWORK ID          NAME                DRIVER              SCOPE
b3bd878057c3        bridge              bridge              local
04893c955700        host                host                local
b54470f6cdd2        none                null                local
```

可以看到当前存在 3 种网络。

4.1.2 容器的 4 种网络模式

1. Bridge 网络模式

Docker 容器的 Bridge 网络模式相当于 Vmware 中的 NAT 模式，容器使用独立的 Network Namespace，并连接到 docker0 虚拟网卡（Docker 进程首次启动时会在当前节点上创建一个名为 docker0 的桥设备，并默认配置其使用 172.17.0.0/16 网络，该网络是 Bridge 网络模式的一种实现，也是创建容器时默认使用的网络），通过 docker0 网桥以及 iptables NAT 表配置与宿主机通信。Bridge 网络模式是 Docker 默认的网络设置，此模式会为每一个容器分配 Network Namespace、设置 IP 等，并将一个主机上的 Docker 容器连接到一个虚拟网桥上。其网络模型如图 4-1 所示。

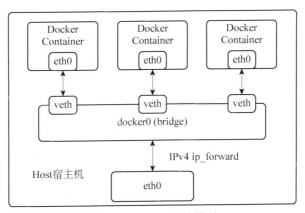

图 4-1 Bridge 网络模型

从上面的网络模型可以看出，容器从原理上来说是可以与宿主机乃至外界的其他机器通信的。同一宿主机上，容器之间都是连接到 docker0 这个网桥上的，它作为虚拟交换机使容器可以相互通信。然而，由于宿主机的 IP 地址与容器 veth pair 的 IP 地址均不在同一个网段，故仅仅依靠 veth pair 和 Namespace 的技术，还不足以使宿主机以外的网络主动发现容器的存在。为了使外界可以访问容器中的进程，Docker 采用了端口绑定的方式，也就是通过 iptables 的 NAT，将宿主机上的端口流量转发到容器内的端口上。

2. Host 网络模式

由于容器和宿主机共享同一个网络命名空间，换言之，容器的 IP 地址即为宿主机的 IP

地址，所以容器可以和宿主机一样，使用宿主机的任意网卡，实现和外界的通信。其网络模型如图 4-2 所示。

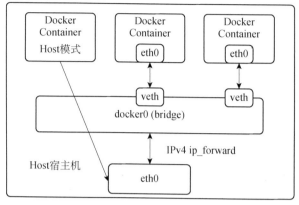

图 4-2　Host 网络模式

Host 网络模式是 Bridge 网络模式很好的补充。采用 Host 网络模式的 Docker 容器，可以直接使用宿主机的 IP 地址与外界进行通信，若宿主机的 eth0 是一个公有 IP，那么容器也拥有这个公有 IP，同时容器内服务的端口也可以使用宿主机的端口，无须额外进行 NAT 转换。当然，有了这样的方便，肯定会损失部分其他的特性，最明显的是 Docker 容器网络环境隔离性的弱化，即容器不再拥有隔离、独立的网络栈。另外，使用 Host 网络模式的 Docker 容器虽然可以让容器内部的服务和传统情况无差别、无改造地被使用，但是由于网络隔离性的弱化，该容器会与宿主机共享并竞争网络栈的使用；此外，容器内部将不再拥有所有的端口资源，原因是部分端口资源已经被宿主机本身的服务占用，还有部分端口已经用以 Bridge 网络模式容器的端口映射（若有其他网络模式的容器运行在该主机上）。

3. None 网络模式

在这种模式下，容器有独立的网络栈，但不包含任何网络配置，只具有 lo 这个 loopback 网卡用于进程通信。也就是说，None 网络模式为容器做了最少的网络设置。实际上，该模式关闭了容器的网络功能。

4. Container 网络模式

Container 网络模式是 Docker 中一种较为特别的网络模式。在这个模式下的容器会使用其他容器的网络命名空间，其网络隔离性会处于 Bridge 网络模式与 Host 网络模式之间。当容器共享其他容器的网络命名空间时，在这两个容器之间不存在网络隔离，而它们又与宿主机以及除此之外其他的容器存在网络隔离。其网络模型如图 4-3 所示。

这种模式下的容器可以通过 localhost 来与同一网络命名空间下的其他容器通信，传输效率较高。而且这种模式还节约了一定数量的网络资源，但它并没有改变容器与外界通信的方式。这在一些特殊的场景中非常有用，例如，Kubernetes 的 Pod，Kubernetes 为 Pod 创建了一个基础设施容器，同一 Pod 下的其他容器都以其他容器模式共享这个基础设施容器的网络命名空间，相互之间以 localhost 访问，构成一个统一的整体。

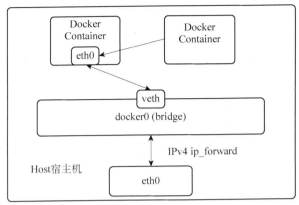

图 4-3 Container 网络模式

4.2 实现容器的多种网络模式

1. 环境准备

（1）规划节点

存储卷节点的规划如表 4-2 所示。

表 4-2 存储卷节点的规划

IP 地址	主机名	节点
192.168.200.14	dockernet	Docker 实验节点

（2）基础准备

准备好 VMware Workstation 工具和 CentOS7.5 虚拟机快照，克隆一台虚拟机，云主机配置自定义，推荐配置 2vCPU、4GB 内存、40GB 硬盘，网络使用 NAT 模式，Docker 节点 IP 设置为 192.168.200.14，推荐设置主机密码为 000000，使用克隆的虚拟机作为实验基础环境。在进行容器网络实验前，自行将虚拟机配置好 IP、安装 Docker 服务，关闭 SELinux 和防火墙服务。

2. 使用 Bridge 网络模式

举个简单的例子来使用 Docker Bridge 网络模式。首先将提供的 nginx_latest.tar 上传至 dockernet 节点，然后加载至本地，命令如下：

```
[root@dockernet ~]# docker load -i nginx_latest.tar
07cab4339852: Loading layer [==================================>]  72.49MB/72.49MB
822ae9fef1d8: Loading layer [==================================>]  64.53MB/64.53MB
7230cfe05cc1: Loading layer [==================================>]  3.072kB/3.072kB
8eb80f066de2: Loading layer [==================================>]  4.096kB/4.096kB
8032102adebe: Loading layer [==================================>]  3.584kB/3.584kB
Loaded image: nginx:latest
```

启动 Nginx 容器，使用 Bridge 网络模式，命令如下：

```
[root@dockernet ~]# docker run -itd --name web -p 80:80 nginx:latest
85eb382d748adc148b382fe0373c4973a8ac3c8da50c8684d5cdf4618f66cdac
```

因为 Bridge 网络模式为容器的默认网络模式,所以在启动时不加任何参数即默认创建为 Bridge 网络模式的容器。使用命令查看容器 IP,命令如下:

```
[root@dockernet ~]# docker inspect -f {{.NetworkSettings.IPAddress}} 85eb3
82d748a
    172.17.0.2
```

查询到 Nginx 容器的 IP 为 172.17.0.2。如果要访问 Nginx 界面,则可以通过宿主机 IP:80 端口来访问,因为在启动时使用了-p 参数,将容器内部的 80 端口与宿主机的 80 端口绑定了。访问界面如图 4-4 所示。

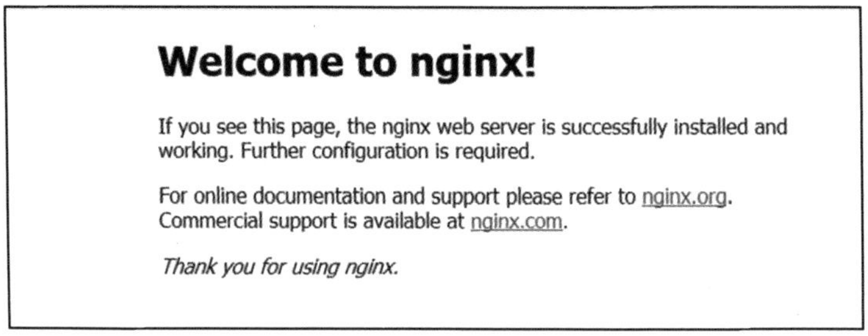

Welcome to nginx!

If you see this page, the nginx web server is successfully installed and working. Further configuration is required.

For online documentation and support please refer to nginx.org.
Commercial support is available at nginx.com.

Thank you for using nginx.

图 4-4 Nginx 首页

在宿主机上,可以通过命令查询 DNAT 规则,具体命令如下:

```
[root@dockernet ~]# iptables -t nat -L -n
Chain PREROUTING (policy ACCEPT)
target     prot opt source               destination
DOCKER     all  --  0.0.0.0/0            0.0.0.0/0            ADDRTYPE match dst-type LOCAL

Chain INPUT (policy ACCEPT)
target     prot opt source               destination

Chain OUTPUT (policy ACCEPT)
target     prot opt source               destination
DOCKER     all  --  0.0.0.0/0            !127.0.0.0/8         ADDRTYPE match dst-type LOCAL

Chain POSTROUTING (policy ACCEPT)
target     prot opt source               destination
MASQUERADE  all  --  172.17.0.0/16       0.0.0.0/0
MASQUERADE  tcp  --  172.17.0.2          172.17.0.2           tcp dpt:80

Chain DOCKER (2 references)
target     prot opt source               destination
```

```
RETURN      all  --  0.0.0.0/0         0.0.0.0/0
DNAT        tcp  --  0.0.0.0/0         0.0.0.0/0         tcp dpt:80
to:172.17.0.2:80
```

很明显，使用 Bridge 网络模式的容器与外界通信时，必定会占用宿主机上的端口，从而与宿主机竞争端口资源，这对宿主机端口的管理会是一个比较大的问题。同时，由于容器与外界通信是基于三层上 iptables NAT 的，其性能和效率上的损耗是可以预见的。

3. 使用 Host 网络模式

使用 Host 网络模式启动云主机，首先将提供的 nginx_latest.tar 镜像上传到 dockernet 节点，并加载到本地，命令如下：

```
[root@dockernet ~]# docker load -i nginx_latest.tar
07cab4339852: Loading layer [==================>]  72.49MB/72.49MB
822ae9fef1d8: Loading layer [==================>]  64.53MB/64.53MB
7230cfe05cc1: Loading layer [==================>]  3.072kB/3.072kB
8eb80f066de2: Loading layer [==================>]  4.096kB/4.096kB
8032102adebe: Loading layer [==================>]  3.584kB/3.584kB
Loaded image: nginx:latest
```

使用 Host 网络模式启动容器，命令如下：

```
[root@dockernet ~]# docker run -itd --net host nginx:latest
eb93f8c02989a439e7b17497e9dafa8596f455ede5e973e4a070701ecb2ff7f7
```

查看容器的运行状态，命令如下：

```
[root@dockernet ~]# docker ps -a
CONTAINER ID        IMAGE               COMMAND                  CREATED
STATUS                  PORTS               NAMES
8912ecc63e68        nginx:latest        "/docker-entrypoint.…"   6 seconds
ago     Exited (1) 3 seconds ago                        admiring_nightingale
dde1959b146a        nginx:latest        "/docker-entrypoint.…"   20 seconds
ago     Up 19 seconds           0.0.0.0:80->80/tcp  web
```

可以看到该容器没有启动，处于 Exited 的状态，原因为：在启动 Nginx 容器的时候使用的是 Host 网络模式，这种模式下会使用宿主机的 IP 和端口，Nginx 容器会默认使用 80 端口。查看宿主机的端口使用情况，如下所示（netstat 命令无法使用的话可以自行使用 yum 命令安装 net-tools 工具）：

```
[root@dockernet ~]# netstat -ntpl
Active Internet connections (only servers)
Proto Recv-Q Send-Q Local Address      Foreign Address     State       PID/Program name
tcp        0      0 0.0.0.0:22         0.0.0.0:*           LISTEN      966/sshd
tcp        0      0 127.0.0.1:25       0.0.0.0:*           LISTEN      1372/master
tcp6       0      0 :::80              :::*                LISTEN      23818/docker-proxy
```

```
tcp6       0      :::22                :::*           LISTEN      966/sshd
tcp6       0      ::1:25               :::*           LISTEN      1372/master
```

可以看见宿主机的 80 端口已被占用，这是因为刚才使用 Bridge 网络模式启动 Nginx 容器时占用了 80 端口，导致 Host 网络模式的容器没有启动。接下来删除 Bridge 网络模式和未启动的两个容器，重新使用 Host 网络模式启动 Nginx 容器，命令如下：

```
[root@dockernet ~]# docker rm -f dde1959b146a 8912ecc63e68
dde1959b146a
8912ecc63e68
[root@dockernet ~]# docker run -itd --net host nginx:latest
651205b42acb11d32f613a7116e0f01c6468dc8ac8a1b5aec397656230d8af10
```

查看容器的运行信息，命令如下：

```
[root@dockernet ~]# docker ps -a
CONTAINER ID     IMAGE           COMMAND                 CREATED       STATUS          PORTS          NAMES
651205b42acb     nginx:latest    "/docker-entrypoint.…"  40 seconds ago   Up 39 seconds                  youthful_blackwell
```

可以看到容器正常启动了，但是没有 PORTS 信息，因为启动容器时使用的是 Host 网络模式，所以容器使用的是宿主机的网络和端口。可以通过访问宿主机的 IP 地址来验证，访问界面如图 4-4 所示。

也可以通过查看宿主机的端口使用情况，来确认容器和宿主机之间的网络关系，如下所示：

```
[root@dockernet ~]# netstat -ntpl
Active Internet connections (only servers)
Proto Recv-Q Send-Q Local Address      Foreign Address      State       PID/Program name
tcp       0      0 0.0.0.0:80         0.0.0.0:*           LISTEN      24032/nginx:master
tcp       0      0 0.0.0.0:22         0.0.0.0:*           LISTEN      966/sshd
tcp       0      0 127.0.0.1:25       0.0.0.0:*           LISTEN      1372/master
tcp6      0      0 :::80              :::*                LISTEN      24032/nginx:master
tcp6      0      0 :::22              :::*                LISTEN      66/sshd
tcp6      0      0 ::1:25             :::*                LISTEN      1372/master
```

可以看到宿主机的 80 端口被 Nginx 进程占用了，但是在启动容器的时候并没有加 -p 参数来指定端口映射，因此可以判断使用 Host 网络模式的容器和宿主机共享网络和端口。

4. 使用 None 网络模式

使用 None 网络模式来启动容器，首先将提供的 centos_latest.tar 镜像上传到 dockernet 节点，并加载到本地，命令如下：

```
[root@dockernet ~]# docker load -i centos_latest.tar
2653d992f4ef: Loading layer [==================>]  216.5MB/216.5MB
```

```
Loaded image: centos:latest
[root@dockernet ~]# docker images
REPOSITORY      TAG         IMAGE ID        CREATED         SIZE
centos          latest      300e315adb2f    3 months ago    209MB
nginx           latest      992e3b7be046    5 months ago    133MB
```

在启动容器时使用 --net none 这个参数来设置容器使用 None 网络模式，命令如下：

```
[root@dockernet ~]# docker run -itd --net none centos:latest
5ceeccbc6d32d2b5b68426bed2dfb612c967e1d1fc83d4de39ff232885826158
```

进入容器查看网络，命令如下：

```
[root@dockernet ~]# docker exec -it 5ceeccbc6d32 /bin/bash
[root@5ceeccbc6d32 /]# ip a
1: lo: <LOOPBACK,UP,LOWER_UP> mtu 65536 qdisc noqueue state UNKNOWN group default qlen 1000
    link/loopback 00:00:00:00:00:00 brd 00:00:00:00:00:00
    inet 127.0.0.1/8 scope host lo
       valid_lft forever preferred_lft forever
```

可以看见该容器的网络只有一个 lo 设备，没有其他网络设备，该容器也不能和外部进行通信。

5. 使用 Container 网络模式

使用 Container 网络模式创建容器，首先将提供的 centos:latest 上传到 dockernet 节点的 /root 目录下，然后加载到本地并查看，命令如下：

```
[root@dockernet ~]# docker load -i centos_latest.tar
2653d992f4ef: Loading layer [==================>]  216.5MB/216.5MB
Loaded image: centos:latest
[root@dockernet ~]# docker images
REPOSITORY      TAG         IMAGE ID        CREATED         SIZE
centos          latest      300e315adb2f    3 months ago    209MB
nginx           latest      992e3b7be046    5 months ago    133MB
```

使用 centos:latest 镜像先创建一个容器 vm1，命令如下：

```
[root@dockernet ~]# docker run -itd --name vm1 centos:latest
e24b8df92c96e76ea40c69a3e19564a9d3b74d2e6d977cdf7db5a5c207e78ace
```

创建容器 vm2，使用 Container 网络模式连接容器 vm1 的网络，命令如下：

```
[root@dockernet ~]# docker run -itd --name vm2 --net container:vm1 centos:latest
cd202b56ef164cced315d22d5cb4af78298034151691df1ecd0a09a72a757a69
```

进入两个容器的内部查看网络，首先进入第一个容器 vm1 查看，命令如下：

```
[root@dockernet ~]# docker exec -it vm1 /bin/bash
[root@e24b8df92c96 /]# ip a
```

```
    1: lo: <LOOPBACK,UP,LOWER_UP> mtu 65536 qdisc noqueue state UNKNOWN group
default qlen 1000
       link/loopback 00:00:00:00:00:00 brd 00:00:00:00:00:00
       inet 127.0.0.1/8 scope host lo
          valid_lft forever preferred_lft forever
    12: eth0@if13: <BROADCAST,MULTICAST,UP,LOWER_UP> mtu 1500 qdisc noqueue
state UP group default
       link/ether 02:42:ac:11:00:02 brd ff:ff:ff:ff:ff:ff link-netnsid 0
       inet 172.17.0.2/16 brd 172.17.255.255 scope global eth0
          valid_lft forever preferred_lft forever
```

再进入第二个容器 vm2 查看，命令如下：

```
[root@dockernet ~]# docker exec -it vm2 /bin/bash
[root@e24b8df92c96 /]# ip a
    1: lo: <LOOPBACK,UP,LOWER_UP> mtu 65536 qdisc noqueue state UNKNOWN group
default qlen 1000
       link/loopback 00:00:00:00:00:00 brd 00:00:00:00:00:00
       inet 127.0.0.1/8 scope host lo
          valid_lft forever preferred_lft forever
    12: eth0@if13: <BROADCAST,MULTICAST,UP,LOWER_UP> mtu 1500 qdisc noqueue
state UP group default
       link/ether 02:42:ac:11:00:02 brd ff:ff:ff:ff:ff:ff link-netnsid 0
       inet 172.17.0.2/16 brd 172.17.255.255 scope global eth0
          valid_lft forever preferred_lft forever
```

可以看见两个容器的 IP 地址一样，vm2 容器使用了 vm1 容器的 IP 地址。Container 网络模式指定新创建的容器和已经存在的容器共享一个 Network Namespace，而不是和宿主机共享。新创建的容器不会创建自己的网卡、配置自己的 IP，而是和一个指定的容器共享 IP、端口范围等。

归纳总结

通过本单元内容的学习，读者应该掌握了容器的 4 种网络模式，也了解了这几种网络模式不同的特点与使用场景。通过实操案例的学习，读者也掌握了使用不同种类网络模式创建容器的方法。更多有关容器网络的知识，读者可以自行查找资料学习。

课后练习

一、判断题

1. Docker 删除正在运行的容器命令是 docker rm -f。（　　）

2. Docker 创建一个新的容器但不启动的命令是 docker run。（　　）

二、单项选择题

1. 查询 Docker 系统信息的命令是（　　）。

   A. docker start　　　B. docker info　　　C. docker version　　　D. docker run

2. Docker 删除一个或者多个镜像的命令是（　　）。

   A. docker rm　　　B. docker del　　　C. docker rmi　　　D. docker run

三、多项选择题

1. 下列关于 Docker 文件系统的说法中错误的是（　　）。

   A. Bootfs 可以修改

   B. Rootfs 不可以修改

   C. 在镜像层新增加一个文件层，则原来的文件会被覆盖

   D. 上层文件层不会覆盖下层文件层中相同的文件

2. 关于 Docker 虚拟化，下列说法中正确的是（　　）。

   A. Docker 基于 Linux 64bits，无法在 32bits 的 Linux/Windows/UNIX 环境下使用

   B. Docker 虚拟化可以替代其他所有虚拟化

   C. Docker 技术可以不基于 OS 系统

   D. Docker 可以在 Windows 上进行虚拟化

技能训练

1. 使用 Host 网络模式创建 Nginx 容器服务。
2. 使用 Bridge 网络模式创建 Nginx 容器服务。

单元 5　容器镜像

学习目标

通过本单元的学习，要求了解 Docker 镜像的概念及系统架构、Docker Commit 的概念和 Dockerfile 的区别、Commit 的操作方法和 Dockerfile 的编写方式。本单元培养读者掌握使用 Docker Commit 构建镜像、使用 Dockerfile 构建镜像等技能；也培养读者谦虚好学、善于思考、善于交流的能力。

5.1　制作容器镜像的方式

1. Docker 镜像概念

Docker 镜像是一个只读的 Docker 容器模板，也是一个独立的文件系统，其中包含了启动 Docker 容器时所需要的文件系统结构及其内容，因此镜像也是启动 Docker 容器的基础。Docker 镜像的文件内容以及一些运行 Docker 容器的配置文件共同组成了 Docker 容器的静态文件系统运行环境 Rootfs，也可以把它理解为 Docker 镜像是 Docker 容器的静态视角，Docker 容器是 Docker 镜像的运行状态。

2. 镜像系统结构

关于 Docker 镜像的系统结构，如图 5-1 所示。

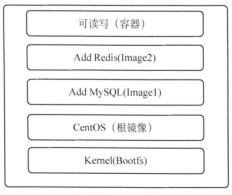

图 5-1　镜像架构

镜像的底层必须是一个称为启动文件系统（Bootfs）的镜像，用户不会直接与这一层打交道。Bootfs 的上层镜像叫作根镜像（Rootfs），在通常情况下它是一个操作系统，例如 Ubuntu、Debian 等。用户的镜像必须构建在根镜像之上，图 5-1 所示的 Image1 是通过在根镜像 CentOS 上安装 MySQL 来创建的，在 Image1 的基础上再安装一个 Redis，就接着创建了 Image2。利用 Image2 启动的容器里面运行的是一个已经安装好了 MySQL 和 Nginx 的 CentOS 系统。

3. Dcoker Commit 介绍

在 Docker 中除了传统的 Linux 容器技术之外，还有其他的镜像技术。镜像技术的采用，使得 Docker 能够自下而上地打包一个完整的应用，能够将更多的精力专注于应用的本身。

Docker 镜像与 Docker 容器之间的关系相辅相成，它们共同作为技术基础来支撑 Docker 的发展，并为 Docker 的生态带来更大的凝聚力。然而，这两项技术并不是相互孤立的，两者之间的相互转换使得 Docker 技术变得尤为方便，说到 Docker 镜像与 Docker 容器之间的相互转换，则要从两个角度来看待：一是从 Docker 镜像转化为 Docker 容器，它们之间一般通过 docker run 命令转化；二是从 Docker 容器转化为 Docker 镜像，这种方式的转换则完全依靠 docker commit 来实现。

通俗地讲，docker commit 命令相当于给运行的容器做快照，类比 OpenStack 云计算平台，容器就是运行的云主机，容器镜像就是虚拟机镜像。在 OpenStack 云平台中，如果需要制作一个 MySQL 镜像，则会启动一个虚拟机，安装 MySQL 服务并初始化，设置开机自启，然后将这台虚拟机打成快照，这个快照也就是一个虚拟机镜像，以后如果要启动带有 MySQL 服务的虚拟机，就选择这个镜像。而 docker commit 命令相当于给运行的容器打快照，并制作成镜像。比如也需要一个 MySQL 镜像，此时需启动一个基础的操作系统容器，安装 MySQL 服务并初始化，设置开机自启，然后使用 docker commit 命令打成镜像，那么今后使用这个镜像启动的容器就带有 MySQL 服务了。

4. Dockerfile 介绍

Dockerfile 是一个包含组合镜像命令的文本文档，可以用在命令行中调用任何命令。Docker 通过读取 Dockerfile 中的指令自动生成镜像。

在 Dockerfile 中，docker build 命令用于从 Dockerfile 中构建镜像。可以在 docker build 命令中使用-f 参数指向文件系统中任何位置的 Dockerfile，语法如下所示：

```
# docker build -f /path/to/a/Dockerfile
```

Dockerfile 一般分为 4 个部分：基础镜像信息、维护者信息、镜像操作指令和容器启动时执行指令，"#"为 Dockerfile 中的注释。

Dockerfile 的主要指令如下：

① FROM：指定基础镜像，这个参数必须为第一条命令。

② MAINTAINER：添加维护者信息。

③ RUN：构建镜像时执行的命令。

④ ADD：将本地文件添加到容器中，tar 类型文件会自动解压（网络压缩资源不会被解压），可以访问网络资源，类似 wget。

⑤ COPY：功能类似 ADD，但是不会自动解压文件，也不能访问网络资源。
⑥ CMD：构建容器后调用，也就是在容器启动时才进行调用。
⑦ ENTRYPOINT：配置容器，使其可执行化。配合 CMD 时可省去"application"，只使用参数。
⑧ LABEL：用于为镜像添加元数据。
⑨ ENV：设置环境变量。
⑩ EXPOSE：指定与外界交互的端口。
⑪ VOLUME：用于指定持久化目录。
⑫ WORKDIR：工作目录，类似于 cd 命令。
⑬ USER：指定运行容器时的用户名或 UID，后续的 RUN 也会使用指定用户。使用 USER 指定用户时，可以使用用户名、UID 或 GID，或是两者的组合。当服务不需要管理员权限时，可通过该命令指定运行用户。
⑭ ARG：用于指定传递给构建运行时的变量。
⑮ ONBUILD：用于设置镜像触发器。

5.2　容器镜像制作

（1）规划节点

制作容器镜像节点的规划如表 5-1 所示。

表 5-1　制作容器镜像节点的规划

IP 地址	主机名	节点
192.168.200.15	docker	Docker 实验节点

（2）基础准备

准备好 VMware Workstation 工具和 CentOS7.5 版本虚拟机，使用 CentOS7.5 虚拟机快照，克隆一台虚拟机，云主机配置自定义，推荐配置 2vCPU、4GB 内存、40GB 硬盘，网络使用 NAT 模式，docker 节点 IP 设置为 192.168.200.15，推荐设置主机密码为 000000。与 4.2 节类似，在进行容器网络实验前，自行将虚拟机配置好 IP、安装 Docker 服务，关闭 SELinux 和防火墙服务。

1. 使用 docker commit 构建镜像

docker commit 命令用于在容器中创建一个新的镜像，其语法如下：

```
# docker commit [OPTIONS] CONTAINER [REPOSITORY[:TAG]]
```

参数如下：
① -a：提交的镜像作者。
② -c：使用 Dockerfile 指令来创建镜像。
③ -m：提交时的说明文字。
④ -p：在 commit 时，将容器暂停。

使用 docker commit 的方式，在基础容器 CentOS 中安装 Java JDK1.8 环境，然后将容器打包成镜像，步骤如下。

（1）上传基础镜像

将提供的 centos_latest.tar 镜像上传至 docker 节点的/root 目录下，然后上传至本地，命令如下：

```
[root@docker ~]# docker load -i centos_latest.tar
2653d992f4ef: Loading layer [================>]  216.5MB/216.5MB
Loaded image: centos:latest
```

（2）启动容器

使用上传的镜像运行容器，命令如下：

```
[root@docker ~]# docker run -itd centos:latest /bin/bash
55f9b5acba8c282d9dc6bd245da24eb23816f0e775e90184ff9743d55e9bf019
```

如果没有报错，即成功启动容器。

（3）上传 JDK 包

首先将提供的 jdk-8u141-linux-x64.tar.gz 包上传至宿主机 docker 节点的/root 目录下，接着将该 JDK 包传至 CentOS 容器的/root 目录下，命令如下：

```
[root@docker ~]# docker cp jdk-8u141-linux-x64.tar.gz 2f0155e8ee9e:/root/
```

然后进入容器内，查看 JDK 包是否被上传，命令如下：

```
[root@docker ~]# docker exec -it 2f0155e8ee9e /bin/bash
[root@2f0155e8ee9e /]# cd /root/
[root@2f0155e8ee9e ~]# ls
anaconda-ks.cfg  anaconda-post.log  jdk-8u141-linux-x64.tar.gz  original-ks.cfg
```

可以看到在容器的/root 目录下存在 jdk-8u141-linux-x64.tar.gz 包，上传成功。

（4）安装 JDK 环境

首先创建 Java JDK 工作目录，命令如下：

```
[root@2f0155e8ee9e ~]# mkdir /usr/jdk64/
```

然后将 jdk-8u141-linux-x64.tar.gz 包解压至/usr/jdk64/目录，命令如下：

```
[root@2f0155e8ee9e ~]# tar -zxvf jdk-8u141-linux-x64.tar.gz -C /usr/jdk64/
......
忽略输出
......
```

解压完毕后，编辑/etc/profile 文件配置环境变量，命令如下：

```
[root@2f0155e8ee9e ~]# vi /etc/profile
#在/etc/profile 文件最下处，添加如下两行
export JAVA_HOME=/usr/jdk64/jdk1.8.0_141
```

```
export PATH=$JAVA_HOME/bin:$PATH
```

编辑环境变量完毕后,需要生效环境变量,命令如下:

```
[root@2f0155e8ee9e ~]# source /etc/profile
```

环境变量生效后,可以使用命令查看 JDK 环境是否部署成功,命令如下:

```
[root@2f0155e8ee9e ~]# java -version
java version "1.8.0_141"
Java(TM) SE Runtime Environment (build 1.8.0_141-b15)
Java HotSpot(TM) 64-Bit Server VM (build 25.141-b15, mixed mode)
```

看到 JDK 版本,即表明环境部署成功。此时虽然 JDK 环境配置成功,但是退出容器重新进入后,Java 命令不能使用,需要重新生效环境变量才行,因此还需要编辑.bashrc 文件,命令如下:

```
[root@2f0155e8ee9e /]# vi ~/.bashrc
#将该文件配置成如下:
if [ -f /etc/bashrc ]; then
    . /etc/bashrc
source /etc/profile
fi
```

配置完之后,保存文件。配置完该文件后,退出容器后重新进入就不需要重复 source 环境变量了,JDK 环境可用。

(5) 将容器做成镜像

在容器中部署完 JDK 环境后,使用 docker commit 命令将容器制作成镜像,并命名为 centos_jdk:1.8,命令如下:

```
[root@docker ~]# docker commit 2f0155e8ee9e centos_jdk:1.8
sha256:4fa97e4a6d00046a08810dd50161a58bec07d83f02034213c529e0754bdd6c26
```

查看镜像列表,命令如下:

```
[root@docker ~]# docker images
REPOSITORY      TAG      IMAGE ID       CREATED         SIZE
centos_jdk      1.8      4fa97e4a6d00   3 seconds ago   771MB
centos          latest   300e315adb2f   9 months ago    209MB
```

可以看到执行 docker commit 命令后,会生成新的镜像,此镜像可以看作是之前容器的快照,使用该镜像运行的容器均自带 JDK 环境。

(6) 验证镜像

首先使用生成的镜像运行容器,然后进入容器内部查看是否带有 JDK 环境,命令如下:

```
[root@docker ~]# docker run -itd centos_jdk:1.8 /bin/bash
384ebb854046bf21fac81aeef61ff268f149e5ac69be52972026aa2686ba811c
[root@docker ~]# docker exec -it 384ebb85404 /bin/bash
[root@384ebb854046 /]# java -version
```

```
java version "1.8.0_141"
Java(TM) SE Runtime Environment (build 1.8.0_141-b15)
Java HotSpot(TM) 64-Bit Server VM (build 25.141-b15, mixed mode)
[root@384ebb854046 /]#
```

可以发现当前容器存在 JDK 环境，证明使用 docker commit 命令制作镜像成功。不过，在日常的工作中，制作镜像一般不使用 docker commit 这种方式，因为使用这种方式制作的镜像不易维护，也就是说其他人不知道作者在这个镜像中做了什么操作，出现问题之后，也不知道从何排查。

在日常工作中，工程师更倾向于使用 Dockerfile 的方式制作镜像，这种方式虽然难度较大，但是镜像结构清晰，易于维护。

2. 使用 Dockerfile 构建镜像

本案例以 centos:latest 为基础镜像，安装 JDK1.8 并构建新的镜像 centos-jdk1.8。

首先将提供的 centos:latest 镜像包以及 JDK1.8 上传至系统，并使用命令将 centos 镜像包导入，命令如下：

```
[root@docker ~]# docker load -i centos_latest.tar
9e607bb861a7: Loading layer [================>]  227.4MB/227.4MB
Loaded image: centos:latest
[root@docker ~]# docker images
REPOSITORY      TAG         IMAGE ID        CREATED         SIZE
centos          latest      0f3e07c0138f    23 months ago   220MB
```

在 docker 宿主机节点中新建一个名为 centos-jdk 的文件夹，用于存放 JDK 的压缩包和 Dockerfile 文件，命令如下：

```
[root@docker ~]# mkdir centos-jdk
[root@docker ~]# mv jdk-8u141-linux-x64.tar.gz  centos-jdk/
[root@docker ~]# cd centos-jdk/
```

编写 Dockerfile 文件，内容如下：

```
[root@docker centos-jdk]# vi Dockerfile
# CentOS with JDK 8
# Author    Jnzz
# 指定基础镜像
FROM centos
# 指定作者
MAINTAINER Jnzz
# 新建文件夹用于存放 JDK 文件
RUN mkdir /usr/local/java
# 将 JDK 文件复制到镜像内并自动解压
ADD jdk-8u141-linux-x64.tar.gz /usr/local/java/
# 创建软链接
RUN ln -s /usr/local/java/jdk1.8.0_141 /usr/local/java/jdk
```

```
# 设置环境变量
ENV JAVA_HOME /usr/local/java/jdk
ENV JRE_HOME ${JAVA_HOME}/jre
ENV CLASSPATH .:${JAVA_HOME}/lib:${JRE_HOME}/lib
ENV PATH ${JAVA_HOME}/bin:$PATH
```

编辑 Dockerfile 之后使用命令开始构建新的镜像，命令如下：

```
[root@docker centos-jdk]# docker build -t centos_jdk2:1.8 .
Sending build context to Docker daemon  185.5MB
Step 1/9 : FROM centos
 ---> 0f3e07c0138f
Step 2/9 : MAINTAINER Jnzz
 ---> Running in aac2a60b125d
Removing intermediate container aac2a60b125d
 ---> bd40ed07bccd
Step 3/9 : RUN mkdir /usr/local/java
 ---> Running in 5b024f865129
Removing intermediate container 5b024f865129
 ---> 0bb0149af865
Step 4/9 : ADD jdk-8u141-linux-x64.tar.gz /usr/local/java/
 ---> 11284fc38cf3
Step 5/9 : RUN ln -s /usr/local/java/jdk1.8.0_141 /usr/local/java/jdk
 ---> Running in 4e1d0301cdb0
Removing intermediate container 4e1d0301cdb0
 ---> b8d2326b9fce
Step 6/9 : ENV JAVA_HOME /usr/local/java/jdk
 ---> Running in 376066d87a1e
Removing intermediate container 376066d87a1e
 ---> 5f0ec8d7f402
Step 7/9 : ENV JRE_HOME ${JAVA_HOME}/jre
 ---> Running in f02ffc89f512
Removing intermediate container f02ffc89f512
 ---> 81ffc9536dbe
Step 8/9 : ENV CLASSPATH .:${JAVA_HOME}/lib:${JRE_HOME}/lib
 ---> Running in 1470e8b69d59
Removing intermediate container 1470e8b69d59
 ---> 3d548c9aefcf
Step 9/9 : ENV PATH ${JAVA_HOME}/bin:$PATH
 ---> Running in 4354649ff767
Removing intermediate container 4354649ff767
 ---> cf7820f2a5a3
Successfully built cf7820f2a5a3
Successfully tagged centos-jdk:latest
```

从上述过程中，可以看出镜像构建历经 9 个步骤，这 9 个步骤在 Dockerfile 文件中按指定的命令及参数运行信息。构建成功后查看镜像，命令如下：

```
[root@docker centos-jdk]# docker images
REPOSITORY      TAG       IMAGE ID        CREATED           SIZE
centos_jdk2     1.8       bf56b57d850c    14 seconds ago    586MB
centos_jdk      1.8       4fa97e4a6d00    34 minutes ago    771MB
centos          latest    0f3e07c0138f    23 months ago     220MB
```

接下来使用命令将构建好的镜像运行为一个容器，并验证 JDK 环境是否被成功地安装，命令如下：

```
[root@docker centos-jdk]# docker run -it centos_jdk2:1.8 /bin/bash
[root@4b7e44d31fe8 /]# java -version
java version "1.8.0_141"
Java(TM) SE Runtime Environment (build 1.8.0_141-b15)
Java HotSpot(TM) 64-Bit Server VM (build 25.141-b15, mixed mode)
```

可以看到 JDK 环境正常运行，使用 Dockerfile 制作镜像成功。用 Dockerfile 方式制作镜像是 Docker 工作者最常用的方式，使用这种方式可以方便镜像的交流，因为只需要把 Dockerfile 文件共享，就可以制作镜像。同时，使用该方式也便于镜像的维护，Dockerfile 中定义了制作镜像的每一步操作，若镜像出现问题，就可以根据 Dockerfile 中的步骤，快速定位问题。

归纳总结

通过本单元内容的学习，读者掌握了使用 Docker Commit 和 Dockerfile 两种方式制作属于自己的容器镜像，同时也了解到 Commit 方式的局限性与不易维护性。在日常的工作中，使用 Dockerfile 方式进行镜像的制作是最常见的，使用 Dockerfile 方式不仅可以让别人知道作者在镜像中做了什么，同时也更易于维护，用户只需要读懂 Dockerfile 文件中的内容，即可自定义修改。

课后练习

一、判断题

1. 在 Docker 的描述文件 Dockerfile 中，FROM 的含义是定义基础镜像。（ ）
2. 在 Docker 的描述文件 Dockerfile 中，MAINTAINER 的含义是作者或者维护者。（ ）

二、单项选择题

1. 在 Docker 的描述文件 Dockerfile 中，RUN 的含义是（ ）。
   A. 定义基础镜像   B. 作者或者维护者
   C. 运行的 Linux 命令   D. 增加文件或目录
2. 在 Docker 的描述文件 Dockerfile 中，ADD 的含义是（ ）。

A. 定义基础镜像　　　　　　　　　　B. 作者或者维护者
C. 运行的 Linux 命令　　　　　　　　D. 增加文件或目录

三、多项选择题

1. 下面关于 Docker 的说法中正确的是（　　）。
A. 创建 Docker 镜像有两种方式，分别是使用文件"dockerfile"和命令"docker commit"。
B. 构建 Docker 镜像时，使用"START"指令指定一个基本镜像。
C. 构建 Docker 镜像时，使用"COPY"指令从用户的工作目录复制文件到镜像中。
D. 构建 Docker 镜像时，使用"WORKDIR"指令指定启动容器时使用的目录。

2. 在 Docker 的描述文件 Dockerfile 中，关于 CMD 的含义下列说法中错误的是（　　）。
A. 定义基础镜像　　　　　　　　　　B. 定义环境变量
C. 运行进程　　　　　　　　　　　　D. 增加文件或目录

技能训练

1. 在容器内安装 JDK 环境，并使用 Docker Commit 的方式构建镜像。
2. 在容器内安装 JDK 环境，并使用 Dockerfile 的方式构建镜像。

单元 6　Docker Compose 编排服务

学习目标

通过本单元的学习，要求了解 Docker Compose 容器编排工具的基本概念与特点、Docker Compose 的安装部署、Docker Compose 的使用方法与技巧等。本单元培养读者掌握 Docker Compose 的安装与使用、Dockerfile 文件编写构建镜像、编排部署 gpmall 商城服务等技能；也培养读者善于交流总结和团队协作的能力。

6.1　认识 Docker Compose 编排服务

1. Docker Compose 简介

Docker Compose 项目是 Docker 官方的开源项目，实现对 Docker 容器集群的快速编排。

Docker Compose 将所管理的容器分为 3 层，分别是工程（Project）、服务（Service）以及容器（Container）。Docker Compose 运行目录下的所有文件（docker-compose.yml、extends 文件或环境变量文件等）组成一个工程，若无特殊指定工程名即为当前目录名。一个工程当中可包含多个服务，每个服务中定义了容器运行的镜像、参数、依赖。一个服务当中可包括多个容器实例，Docker Compose 并没有解决负载均衡的问题，因此需要借助其他工具实现服务发现及负载均衡。

Docker Compose 的工程配置文件默认为 docker-compose.yml，可通过环境变量 COMPOSE_FILE 或 -f 参数自定义配置文件，其定义了多个有依赖关系的服务及每个服务运行的容器。

使用一个 Dockerfile 模板文件，可以让用户很方便地定义一个单独的应用容器。在工作中，经常会碰到需要多个容器相互配合来完成某项任务的情况。

比如网站开发最常见的场景：网站需要有数据库、网站应用、Nginx，它们必须互相配合才算是完整的环境。为了简单起见，完全可以以 CentOS 为基础镜像，把这些服务一股脑全装进去，然后运行。但是这样操作有很多缺点，比如每次都要重新安装 MySQL 而不是直接利用 MySQL 官方的基础镜像，升级维护不方便；如果想要扩展应用也很难，因为每个应用都连接着自己内部的数据库，无法共享数据；事实上，这种方式是典型的虚拟机的使用方式，而不是 Docker 的正确打开方式。

Docker 是轻量化的应用程序，Docker 官方推荐每个 Docker 容器中只运行一个进程，那么就是说，需要分别为应用、数据库、Nginx 创建单独的 Docker 容器，然后分别启动它们。想象一下，安装好 Docker 服务之后，每次启动网站，都要至少使用 docker run 命令 3 次，是不是很烦琐？而且此时这几个服务的容器是分散独立的，管理起来很不方便。既然这几个容器都是为了同一个网站服务的，是不是应该把它们放到一起？这就引出了 Docker Compose 项目。

2. Docker Compose 的基本概念

Docker Compose 有 2 个重要的概念。

① 项目（Project）：由一组关联的应用容器组成的一个完整业务单元，在 docker-compose.yaml 文件中定义。

② 服务（Service）：一个应用的容器，实际上可以包括若干运行相同镜像的容器实例。

3. Docker Compose 的特点

（1）将单个主机隔离成多个环境

Compose 使用项目名称（Project Name）将不同应用的环境隔离开，项目名称可以用来：
① 在开发机上，将应用环境复制多份。
② 防止使用了相同名称服务的应用之间互相干扰。

在默认情况下，项目名称是项目文件夹根目录的名称，用户可以使用-p 标识或 COMPOSE_PROJECT_NAME 改变默认的名称。

（2）保护卷中的数据

Compose 保护服务使用的所有卷（Vloumes），当运行 docker-compose run 命令时，如果 Compose 发现存在之前运行过的容器，它就会把旧容器中的数据卷拷贝到新的容器中，这保证了用户在卷中创建的任何数据都不丢失。

（3）只重新创建改变过的容器

Compose 会缓存用于创建容器的配置信息，当重启服务时，如果服务没有被更改，Compose 就会重新使用已经存在的容器，这无疑加快了修改应用的速度。

Compose 文件是一个 Yaml 文件，用于定义 Services、Network 和 Volumes。Compose 文件的默认路径为./docker-compose.yml（后缀为.yml 或.yaml 都可以）。

一个 Service 配置将会应用到容器的启动中，就像将命令行参数传递给 docker run。同样，在 Yaml 文件中 Network 和 Volume 的定义类似于 docker network create 和 docker volume create。

6.2 使用 Docker Compose 编排服务

1. 环境准备

（1）规划节点

Docker Compose 编排实验节点的规划如表 6-1 所示。

表 6-1　Docker Compose 编排实验节点的规划

IP 地址	主机名	节点
192.168.200.16	docker	Compose 实验节点

（2）基础准备

准备好 VMware Workstation 工具、CentOS7.5 版本虚拟机，使用 CentOS7.5 版本虚拟机快照克隆一台虚拟机，云主机配置自定义，推荐配置 2vCPU、4GB 内存、40GB 硬盘，网络使用 NAT 模式，docker 节点 IP 设置为 192.168.200.16，推荐设置主机密码为 000000。自行将虚拟机配置好 IP、安装 Docker 服务，关闭 SELinux 和防火墙服务。

2. Docker Compose 的安装与使用

（1）Docker Compose 的安装

将提供的 v1.25.5-docker-compose-Linux-x86_64 文件上传至 docker 节点，然后复制该文件至/usr/local/bin/目录下，并改名为 docker-compose，命令如下：

```
[root@docker ~]# cp v1.25.5-docker-compose-Linux-x86_64 /usr/local/bin/docker-compose
```

给 docker-compose 赋予执行权限，命令如下：

```
[root@docker ~]# chmod +x /usr/local/bin/docker-compose
```

赋予权限后，即可使用 docker-compose 命令，查看 docker-compose 的版本，命令如下：

```
[root@docker ~]# docker-compose version
docker-compose version 1.25.5, build 8a1c60f6
docker-py version: 4.1.0
CPython version: 3.7.5
OpenSSL version: OpenSSL 1.1.0l  10 Sep 2019
```

至此，docker-compose 工具安装成功。

（2）Docker Compose 的使用

假设当前要部署一个 ownCloud 应用（ownCloud 是一个开源、免费、专业的私有云存储项目，它能帮用户快速在个人 PC 或服务器上架设一套专属的私有云文件同步网盘），部署 ownCloud 应用需要两个容器，一个是 ownCloud 容器，另一个是数据库容器。本次实验使用两种方式启动 ownCloud 云网盘应用，一种是单容器化部署（即手动部署），另一种是使用 docker-compose 工具编排部署（简称编排方式），读者可以对比这两种方式的不同与优缺点。

方法一：手动部署 ownCloud 云网盘。

首先将提供的 mysql-5.6.tar 镜像和 owncloud-latest.tar 镜像上传至 docker 节点的/root 目录下，并载入，命令如下：

```
[root@docker ~]# docker load -i owncloud-latest.tar
7b4e562e58dc: Loading layer [================>]  58.44MB/58.44MB
5fbcd9a02436: Loading layer [================>]  3.584kB/3.584kB
2fbf086236e5: Loading layer [================>]  213.6MB/213.6MB
c25c4aba0c63: Loading layer [================>]  3.584kB/3.584kB
bcd6f0091bba: Loading layer [================>]  43.04MB/43.04MB
6c26c5566c37: Loading layer [================>]  11.78kB/11.78kB
```

```
c8500dc8d342: Loading layer [==================>]  9.728kB/9.728kB
58cd3b2b073c: Loading layer [==================>]  7.68kB/7.68kB
f4d3eff47c62: Loading layer [==================>]  13.47MB/13.47MB
cd27e9cdba62: Loading layer [==================>]  4.096kB/4.096kB
9ab9eef0e136: Loading layer [==================>]  57.52MB/57.52MB
d3fe85001836: Loading layer [==================>]  11.78kB/11.78kB
708942a92eb6: Loading layer [==================>]  4.608kB/4.608kB
eccae8073817: Loading layer [==================>]  4.608kB/4.608kB
14787a8a491e: Loading layer [==================>]  112.9MB/112.9MB
f6317973ad48: Loading layer [==================>]  3.433MB/3.433MB
14b163b7ec57: Loading layer [==================>]  4.608kB/4.608kB
d62c355d2ddb: Loading layer [==================>]  6.144kB/6.144kB
cf282378a618: Loading layer [==================>]  4.394MB/4.394MB
abfc4688e092: Loading layer [==================>]  130.4MB/130.4MB
1bb18f0308ff: Loading layer [==================>]  3.584kB/3.584kB
Loaded image: owncloud:latest
[root@docker ~]# docker load -i mysql-5.6.tar
767a7b7a8ec5: Loading layer [==================>]  58.49MB/58.49MB
b95d109635ed: Loading layer [==================>]  338.4kB/338.4kB
2ce7750ef4a3: Loading layer [==================>]  10.44MB/10.44MB
f7d565e8883f: Loading layer [==================>]  4.174MB/4.174MB
416da68a865d: Loading layer [==================>]  1.536kB/1.536kB
b660817215fb: Loading layer [==================>]  41.39MB/41.39MB
b36d6fe7a9b6: Loading layer [==================>]  34.3kB/34.3kB
d0eb1caf404c: Loading layer [==================>]  3.584kB/3.584kB
6ed1ec213f77: Loading layer [==================>]  192.6MB/192.6MB
e201e7ba69f1: Loading layer [==================>]  16.38kB/16.38kB
5b1a817757f0: Loading layer [==================>]  1.536kB/1.536kB
Loaded image: mysql:5.6
```

上传完毕后，首先启动数据库容器，命令如下：

```
[root@docker ~]# docker run --name my-mysql -p 3307:3306 -e MYSQL_ROOT_PASSWORD=123456 -d mysql:5.6
37713e86c8af8db1fc9473a18395d82a7b21d7a6d2ffb46a342faf54237cfa2c
```

启动数据库容器，将容器命名为 my-mysql，配置内部数据库的 3306 端口映射到宿主机的 3307 端口，配置数据库密码为 123456。

接着启动 ownCloud 容器，命令如下：

```
[root@docker ~]# docker run --name my-owncloud -d -p 5679:80 -v /data/db/owncloud:/var/www/html/data --link my-mysql:mysql owncloud:latest
c89f5452ae0e5c3390d5ee6646d37d2c3ce1018e13b0e7c385187a74632f8a1c
```

启动 ownCloud 容器，将容器命名为 my-owncloud，映射内部 80 端口到宿主机的 5679 端口；配置数据卷的连接，将内部的/var/www/html/data 目录连接至外部/data/db/owncloud；配置数据库的连接，将数据库连接至上面启动的 my-mysql 数据库。启动成功后，可以通过浏览器访问 http://192.168.200.16:5679 ownCloud 登录界面，如图 6-1 所示。

图 6-1 ownCloud 登录界面

在登录界面上，通过提示，可以进行初始化配置，读者可以自行安装配置，此处不再赘述。使用单容器启动 ownCloud 云网盘时，需要一次次运行容器，而且运行的容器比较分散，不方便管理，接下来使用 docker-compose 工具编排部署 ownCloud 云网盘应用。

方法二：使用 docker-compose 工具编排部署 ownCloud 云网盘。

使用 docker-compose 编排工具，首先需要创建工作目录，在 docker 节点/root 目录下创建 ownCloud 目录，命令如下：

```
[root@docker ~]# mkdir owncloud
```

在 ownCloud 目录下，创建 docker-compose.yaml 文件，文件内容如下：

```yaml
version: '3'
services:
  owncloud:
    image: owncloud:latest
    links:
      - mysql:mysql
    volumes:
      - "/data/db/owncloud:/var/www/html/data"
    ports:
      - 5680:80
  mysql:
    image: mysql:5.6
    volumes:
      - "/data/db/mysql:/var/lib/mysql"
    ports:
      - 3306:3306
```

```
    environment:
      MYSQL_ROOT_PASSWORD: "123456"
      MYSQL_DATABASE: ownCloud
```

编写完 docker-compose.yaml 文件后，保存退出。在该文件中，定义了两个容器使用什么镜像、数据库连接到哪里、数据卷怎么连接、端口怎么映射，还定义了数据库的密码、库名之类的内容。定义好文件之后，要确认端口没有跟现有运行端口冲突、镜像名没有写错，然后可以使用 docker-compose 命令运行该 Yaml 文件，ownCloud 应用会自动编排部署，命令如下：

```
[root@docker owncloud]# docker-compose up -d
Creating network "owncloud_default" with the default driver
Creating owncloud_mysql_1   ... done
Creating owncloud_owncloud_1 ... done
```

看到两个容器都处于启动完成的状态，就可以通过浏览器 http://192.168.200.16:5680 访问 ownCloud 登录界面，参考图 6-1 所示。

使用 docker-compose 方式一样可以运行 ownCloud 应用，而且不需要手动运行容器，只需要运行 docker-compose.yaml 文件。如果需要关闭该容器，手动方式需要关停两次容器，而使用编排方式则只需要执行一条命令，命令如下：

```
[root@docker owncloud]# docker-compose down
Stopping owncloud_owncloud_1 ... done
Stopping owncloud_mysql_1    ... done
Removing owncloud_owncloud_1 ... done
Removing owncloud_mysql_1    ... done
Removing network owncloud_default
```

docker-compose down 命令将停止运行的容器，这个 down 命令就相当于 stop 和 rm 两条命令的结合，它会先停掉容器，然后再删除容器及已创建的所有网络。如果只想将容器停止不想删除的话，可以使用 docker-compose stop 命令。

通过上述两个小案例，可以发现使用 docker-compose 编排工具能够实现对容器的统一管理。ownCloud 网盘应用涉及两个容器，对于读者来说，可能使用手动部署方式与编排方式的差距不大，但是在今后的工作生活中，如果一个应用涉及了 5 个或更多的容器，那么使用编排方式的优势就能体现出来了。在下面的实战案例中，会使用编排方式部署一个应用商城案例，以便读者更深入地体验 docker-compose 编排工具。

3. gpmall 商城应用介绍

在编排部署 gpmall 商城应用之前，首先必须知道该应用涉及哪些服务，然后规划与制作各个容器镜像，最后才能进行编排部署。

首先快速地在虚拟机中安装部署 gpmall 商城应用，了解该应用依赖哪些基础服务。

（1）基础环境安装

gpmall 商城应用需要依赖数据库、Redis、ZooKeeper、Kafka 等基础服务，所以需要安装这些基础服务，具体操作如下：

① 配置 YUM 源。将提供的 gpmall-single.tar.gz 软件包上传至 docker 节点的/root 目录下，然后解压缩到/opt 目录下，命令如下：

```
[root@docker ~]# tar -zxf gpmall-single.tar.gz -C /opt/
```

移除/etc/yum.repos.d/原有的 repo 文件，并新建 local.repo 文件，编辑 local.repo 文件，具体命令如下：

```
[root@docker ~]# mv /etc/yum.repos.d/* /media/
[root@docker ~]# vi /etc/yum.repos.d/local.repo
```

local.repo 文件的内容如下：

```
[gpmall]
name=gpmall
baseurl=file:///opt/gpmall-single/gpmall-repo
gpgcheck=0
enabled=1
```

配置完之后，使用如下命令查看 YUM 源是否配置成功：

```
[root@docker ~]# yum repolist
Loaded plugins: fastestmirror
Determining fastest mirrors
gpmall                                         | 2.9 kB  00:00:00
gpmall/primary_db                              |  98 kB  00:00:00
repo id            repo name              status
gpmall             gpmall                 129
repolist: 129
```

看到 repolist 的值是 129，即表明 YUM 源配置成功。

② 安装 Java 环境。配置完 YUM 源之后，安装基础环境，首先安装 Java 环境，命令如下：

```
[root@docker ~]# yum install -y java-1.8.0-openjdk java-1.8.0-openjdk-devel
...忽略输出...
```

查看 Java 版本，确认 Java 环境是否安装成功，命令如下：

```
[root@docker ~]# java -version
openjdk version "1.8.0_262"
OpenJDK Runtime Environment (build 1.8.0_262-b10)
OpenJDK 64-Bit Server VM (build 25.262-b10, mixed mode)
```

可以看到 Java 版本，说明 Java 安装正确。

③ 安装数据库服务。安装 MariaDB 数据库服务，命令如下：

```
[root@docker ~]# yum install mariadb-server mariadb -y
...忽略输出...
```

④ 安装 Redis 服务。安装 Redis 缓存服务，命令如下：

```
[root@docker ~]# yum install redis -y
...忽略输出...
```

⑤ 安装 Nginx 服务。安装 Nginx 反向代理服务，命令如下：

```
[root@docker ~]# yum install nginx -y
...忽略输出...
```

⑥ 安装 ZooKeeper。安装 ZooKeeper 服务，使用提供的 zookeeper-3.4.14.tar.gz 软件包（该软件包在/opt/gpmall-single 目录内），解压压缩包的命令如下：

```
[root@docker gpmall-single]# tar -zxvf zookeeper-3.4.14.tar.gz
```

进入 zookeeper-3.4.14/conf 目录，将 zoo_sample.cfg 文件重命名为 zoo.cfg，命令如下：

```
[root@docker gpmall-single]# mv zoo_sample.cfg zoo.cfg
```

进入 zookeeper-3.4.14/bin 目录，启动 ZooKeeper 服务，命令如下：

```
[root@docker gpmall-single]# ./zkServer.sh start
ZooKeeper JMX enabled by default
Using config: /root/zookeeper-3.4.14/bin/../conf/zoo.cfg
Starting zookeeper ... STARTED
```

查看 ZooKeeper 状态，命令如下：

```
[root@docker gpmall-single]# ./zkServer.sh status
ZooKeeper JMX enabled by default
Using config: /root/zookeeper-3.4.14/bin/../conf/zoo.cfg
Mode: standalone
```

⑦ 安装 Kafka 服务。安装 Kafka 服务，使用提供的 kafka_2.11-1.1.1.tgz 软件包（该软件包在/opt/gpmall-single 目录内），解压该压缩包，命令如下：

```
[root@docker gpmall-single]# tar -zxvf kafka_2.11-1.1.1.tgz
```

进入 kafka_2.11-1.1.1/bin 目录，启动 Kafka 服务，命令如下：

```
[root@docker gpmall-single]# ./kafka-server-start.sh -daemon ../config/server.properties
```

使用 jps 查看 Kafka 是否成功启动，命令如下：

```
[root@docker gpmall-single]# jps
6039 Kafka
1722 QuorumPeerMain
6126 Jps
```

还可以使用 netstat 命令查看 Kafka 服务端口是否放开，如下所示：

```
[root@docker bin]# netstat -ntpl
```

```
Active Internet connections (only servers)
Proto Recv-Q Send-Q Local Address      Foreign Address    State      PID/Program name
tcp        0      0 0.0.0.0: 22         0.0.0.0:*          LISTEN     1025/sshd
tcp        0      0 127.0.0.1:25        0.0.0.0:*          LISTEN     1292/master
tcp6       0      0 : : :9092           :::*               LISTEN     6164/java
tcp6       0      0 : : :39141          :::*               LISTEN     6164/java
tcp6       0      0 : : :2181           :::*               LISTEN     5856/java
tcp6       0      0 : : :3307           :::*               LISTEN     4457/docker-proxy
tcp6       0      0 : : :5679           :::*               LISTEN     4678/docker-proxy
tcp6       0      0 : : :22             :::*               LISTEN     1025/sshd
tcp6       0      0 : : :25             :::*               LISTEN     1292/master
tcp6       0      0 : : :33211          :::*               LISTEN     5856/java
```

在运行结果中看到了 9092 端口，说明 Kafka 服务已启动。至此，gpmall 商城应用所需要的基础服务均安装完毕。在安装完服务后，还需对各种服务进行配置，具体步骤详见下面的服务配置实操案例。

（2）服务配置

① 数据库服务配置。

需要对安装完的数据库进行初始化配置、设置数据库密码、设置访问权限、导入数据库等，具体操作如下：

首先启动数据服务并进行数据库初始化操作，命令如下：

```
[root@docker ~]# systemctl start mariadb
```

设置 root 用户的密码为 123456 并登录。

```
[root@docker ~]# mysql_secure_installation
/usr/bin/mysql_secure_installation: line 379: find_mysql_client: command not found
NOTE: RUNNING ALL PARTS OF THIS SCRIPT IS RECOMMENDED FOR ALL MariaDB
      SERVERS IN PRODUCTION USE! PLEASE READ EACH STEP CAREFULLY!
In order to log into MariaDB to secure it, we'll need the current
password for the root user. If you've just installed MariaDB, and
you haven't set the root password yet, the password will be blank,
so you should just press enter here.
Enter current password for root (enter for none):     #默认按回车键
OK, successfully used password, moving on...
Setting the root password ensures that nobody can log into the MariaDB
root user without the proper authorisation.
Set root password? [Y/n] y
New password:                                         #输入数据库 root 密码 123456
Re-enter new password:                                #重复输入密码 123456
Password updated successfully!
Reloading privilege tables..
```

```
... Success!
By default, a MariaDB installation has an anonymous user, allowing anyone
to log into MariaDB without having to have a user account created for
them. This is intended only for testing, and to make the installation
go a bit smoother. You should remove them before moving into a
production environment.
Remove anonymous users? [Y/n] y        #输入 y 删除匿名用户
... Success!
Normally, root should only be allowed to connect from 'localhost'. This
ensures that someone cannot guess at the root password from the network.
Disallow root login remotely? [Y/n] n     输入 n,允许 root 远程登录
... skipping.
By default, MariaDB comes with a database named 'test' that anyone can
access. This is also intended only for testing, and should be removed
before moving into a production environment.
Remove test database and access to it? [Y/n] y    输入 y,移除测试库
 - Dropping test database...
... Success!
 - Removing privileges on test database...
... Success!
Reloading the privilege tables will ensure that all changes made so far
will take effect immediately.
Reload privilege tables now? [Y/n] y     输入 y,重新加载授权信息
... Success!
Cleaning up...
All done! If you've completed all of the above steps, your MariaDB
installation should now be secure.
Thanks for using MariaDB!
```

看到 Success 信息,表明初始化数据库成功,接下来设置数据库的远程访问权限,命令如下:

```
[root@mall ~]# mysql -uroot -p123456
Welcome to the MariaDB monitor. Commands end with ; or \g.
Your MariaDB connection id is 9
Server version: 10.3.18-MariaDB MariaDB Server

Copyright (c) 2000, 2018, Oracle, MariaDB Corporation Ab and others.

Type 'help;' or '\h' for help. Type '\c' to clear the current input statement.

MariaDB [(none)]> grant all privileges on *.* to root@localhost identified by '123456' with grant option;
Query OK, 0 rows affected (0.001 sec)
```

```
MariaDB [(none)]> grant all privileges on *.* to root@"%" identified by
'123456' with grant option;
  Query OK, 0 rows affected (0.001 sec)
```

以上两条命令设置了 root 用户的本地和远程登录权限。

最后还需要创建 gpmall 库,并将提供的 gpmall.sql 文件(该文件在/opt/gpmall-single 目录下)导入该库,命令如下:

```
MariaDB [(none)]> create database gpmall;
Query OK, 1 row affected (0.00 sec)
MariaDB [(none)]> use gpmall;
MariaDB [mall]> source /opt/gpmall-single/gpmall.sql
```

导入 gpmall.sql 之后,退出数据库,并设置数据库服务开机自启,命令如下:

```
MariaDB [mall]> Ctrl-C -- exit!
Aborted
[root@mall ~]# systemctl enable mariadb
Created          symlink                                      from
/etc/systemd/system/multi-user.target.wants/mariadb.service                to
/usr/lib/systemd/system/mariadb.service.
```

至此,数据库服务配置完毕。

② Redis 服务配置。Redis 服务的配置比较简单,只需要修改 Redis 的配置文件,取消保护模式和访问限制,具体操作步骤如下:

```
[root@docker ~]# vi /etc/redis.conf
#编辑/etc/redis.conf 配置文件
找到 bind 127.0.0.1 这一行,将这一行注释掉;
找到 protected-mode yes 这一行,将 yes 改为 no。
```

修改完之后,保存退出文件,然后启动 Redis 服务并设置开机自启,命令如下:

```
[root@docker ~]# systemctl start redis
[root@docker ~]# systemctl enable redis
Created symlink from /etc/systemd/system/multi-user.target.wants/redis.
service to /usr/lib/systemd/system/redis.service.
```

这样,Redis 缓存服务配置完毕。

③ Nginx 服务配置。将 dist 目录下的文件(dist 目录在/opt/gpmall-single 目录下)复制到 Nginx 默认项目路径(首先清空默认项目路径下的文件)。

```
[root@docker ~]# rm -rf /usr/share/nginx/html/*
[root@docker ~]# cp -rvf /opt/gpmall-single/dist/* /usr/share/nginx/html/
```

修改 Nginx 配置文件/etc/nginx/conf.d/default.conf,添加映射如下:

```
[root@docker ~]# vi /etc/nginx/conf.d/default.conf
server {
```

```
        listen       80;
        server_name  localhost;

        #charset koi8-r;
        #access_log  /var/log/nginx/host.access.log  main;

        location / {
           root   /usr/share/nginx/html;
           index  index.html index.htm;
        }
     #添加如下 user、shopping、cashier 的访问地址
        location /user {
             proxy_pass http://127.0.0.1:8082;
        }

        location /shopping {
             proxy_pass http://127.0.0.1:8081;
        }

        location /cashier {
             proxy_pass http://127.0.0.1:8083;
        }
        #error_page  404              /404.html;
```

重启 Nginx 服务并设置开机自启,命令如下:

```
[root@docker ~]# systemctl restart nginx
[root@docker ~]# systemctl enable nginx
Created symlink from /etc/systemd/system/multi-user.target.wants/nginx.service to /usr/lib/systemd/system/nginx.service.
```

至此,Nginx 服务部署完毕。

④ 全局变量配置。最后,配置全局变量,修改/etc/hosts 文件,修改项目全局配置文件如下:

```
[root@docker ~]# cat /etc/hosts
127.0.0.1   localhost localhost.localdomain localhost4 localhost4.localdomain4
::1         localhost localhost.localdomain localhost6 localhost6.localdomain6
#添加如下四行
192.168.200.16 kafka.mall
192.168.200.16 mysql.mall
192.168.200.16 redis.mall
192.168.200.16 zookeeper.mall
```

配置完毕后保存并退出文件。完成上述所有配置后,gpmall 商城应用的所有基本服务及配置就准备完毕了,接下来只要启动后端的 jar 包,就可以运行 gpmall 商城应用了。

（3）启动应用

将提供的 4 个 jar 包按照顺序启动，启动命令如下：

```
[root@docker gpmall-single]# nohup java -jar shopping-provider-0.0.1-SNAPSHOT.jar &
[1] 6432
[root@docker gpmall-single]# nohup: ignoring input and appending output to 'nohup.out'

[root@docker gpmall-single]# nohup java -jar user-provider-0.0.1-SNAPSHOT.jar &
[2] 6475
[root@docker gpmall-single]# nohup: ignoring input and appending output to 'nohup.out'

[root@docker gpmall-single]# nohup java -jar gpmall-shopping-0.0.1-SNAPSHOT.jar &
[3] 6523
[root@docker gpmall-single]# nohup: ignoring input and appending output to 'nohup.out'

[root@docker gpmall-single]# nohup java -jar gpmall-user-0.0.1-SNAPSHOT.jar &
[4] 6563
[root@docker gpmall-single]# nohup: ignoring input and appending output to 'nohup.out'
```

按照顺序运行 4 个 jar 包后，后端服务部署完毕，此时可以浏览器访问商城首页 http://192.168.200.16，如图 6-2 所示。

图 6-2　商城首页

4. 容器化构建商城镜像

通过上述实验手动部署 gpmall 商城应用，分析如何分解各依赖服务、制作容器镜像、编排部署商城应用。

按照容器的运行规则，一个容器提供一种服务，在商城应用中，用到了 MariaDB 数据库服务、Redis 缓存服务、Java 环境、ZooKeeper 服务、Kafka 服务、前端 Nginx 服务、后端 jar 包服务。

通过仔细分析，Java 环境是为了运行 jar 包而安装的，Kafka 服务依赖 ZooKeeper 服务，最后确定如果要容器化部署 gpmall 商城应用，需要制作 5 个容器镜像，分别如下：

① MariaDB 镜像提供数据库服务。
② Redis 镜像提供缓存服务。
③ ZooKeeper 镜像提供协调服务。
④ Kafka 镜像提供发布订阅消息服务。
⑤ Nginx 镜像提供前端界面与后端功能服务。

确定了要制作的镜像后，可以根据上面实操的步骤，编写 Dockerfile 文件，制作容器镜像，最后通过 docker-compose 编排工具编排部署 gpmall 商城应用。

（1）制作 Redis 镜像

首先创建制作 Redis 镜像的工作目录，命令如下：

```
[root@docker opt]# mkdir redis
```

此处在 /opt 目录下创建 Redis 工作目录。根据手动安装 Redis 和配置的方式，编写 Dockerfile 文件，文件内容如下：

```
FROM centos:centos7.5.1804
MAINTAINER chinaskill
EXPOSE 6379
RUN rm -rf /etc/yum.repos.d/*
COPY ftp.repo /etc/yum.repos.d/ftp.repo
RUN yum install redis -y
RUN sed -i "s/bind 127.0.0.1/bind 0.0.0.0/g" /etc/redis.conf
RUN sed -i "s/protected-mode.*/protected-mode no/g" /etc/redis.conf
CMD redis-server /etc/redis.conf
```

从 Dockerfile 中可以看到，首先使用的基础镜像是 CentOS7.5 版本，将提供的 centos-centos7.5.1804.tar 镜像包上传至 Docker 节点的 /root 目录下，然后加载，加载完毕后使用如下命令查看：

```
[root@docker ~]# docker images
REPOSITORY      TAG              IMAGE ID       CREATED        SIZE
centos          centos7.5.1804   cf49811e3cdb   2 years ago    200MB
```

然后再看 Dockerfile 文件，作者名叫 chinaskill，这个地方可以自定义，可以修改成自己的名字。

文件暴露了 6379 端口，因为 Redis 服务的默认服务端口是 6379，所以需要将该端口暴露。

删除原有的 repo 文件，为创建自定义 repo 文件做准备。

将 Redis 目录下的 ftp.repo 文件复制到容器内部的 /etc/yum.repos.d/ 目录下，ftp.repo 文

件内容如下：

```
[centos]
name=centos
gpgcheck=0
enabled=1
baseurl=ftp://192.168.200.16/centos
[gpmall]
name=gpmall
gpgcheck=0
enabled=1
baseurl=ftp://192.168.200.16/gpmall-single/gpmall-repo
```

可以看到此处需要在容器内安装 Redis 服务，且使用 FTP 的方式进行安装，所以需要在宿主机 docker 节点安装 FTP 服务，然后将/opt 目录共享，容器内部的 ftp.repo 文件 baseurl 指向宿主机的 FTP 地址，这样容器内部安装 Redis 服务时就可以使用宿主机的 FTP 源了。

关于如何安装 FTP 服务和配置，此处不再赘述（安装完 FTP 服务器后要确保 SELinux 和防火墙处于关闭状态）。还注意到，ftp.repo 文件中还有一个 CentOS 的源，这是因为基础镜像为 CentOS7.5 的容器，在容器中安装服务，可能会出现没有依赖包的情况，所以要把 CentOS 的源挂上。

配置完 YUM 源之后，在 Dockerfile 文件中安装 Redis 服务并配置，最后设置 Redis 服务的启动配置。

梳理完 Dockerfile 文件之后，进行镜像的制作，确认/opt/redis 目录下存在 Dockerfile 文件和 ftp.repo 文件，且基本环境也配置完毕，制作镜像的命令如下：

```
[root@docker redis]# docker build -t gpmall-redis:v1.0 .
Sending build context to Docker daemon  3.072kB
Step 1/9 : FROM centos:centos7.5.1804
 ---> cf49811e3cdb
Step 2/9 : MAINTAINER chinaskill
...忽略输出...
Successfully built 5ab933f160f5
Successfully tagged gpmall-redis:v1.0
```

从 docker build 命令中可以看到制作的镜像名为 gpmall-redis:v1.0，没有报错，最后显示 Successfully built，说明镜像制作成功，可以查看镜像列表，如下所示：

```
[root@docker redis]# docker images
REPOSITORY       TAG      IMAGE ID        CREATED         SIZE
gpmall-redis     v1.0     5ab933f160f5    2 minutes ago   243MB
```

如果制作 Redis 镜像失败，则可以从 YUM 源排查问题，查看宿主机的 FTP 服务是否安装成功，是否把/opt 目录设为访问目录，SELinux 和 Firewalld 防火墙是否关闭，/opt 目录下的文件是否正确，ftp.repo 文件是否准确。

(2)制作数据库镜像

同样地,首先要创建制作镜像的工作目录,命令如下:

[root@docker ~]# mkdir /opt/mariadb

根据手动安装 MariaDB 和配置的方式,编写 Dockerfile 文件,文件内容如下:

```
FROM centos:centos7.5.1804
MAINTAINER chinaskill
EXPOSE 3306
ENV LC_ALL en_US.UTF-8
RUN rm -rf /etc/yum.repos.d/*
COPY ftp.repo /etc/yum.repos.d/ftp.repo
RUN yum install mariadb mariadb-server -y
COPY setup.sh /root/setup.sh
COPY gpmall.sql /root/gpmall.sql
RUN chmod 755 -R /root/setup.sh
RUN /root/setup.sh
CMD ["mysqld_safe"]
```

从 Dockerfile 中可以看到,它同样基于 CentOS7.5 的基础镜像;作者为 chinaskill;暴露 3306 端口;配置数据库的字符编码为 UTF-8;删除容器中原有的 repo 文件;复制 ftp.repo 文件到容器内部的/etc/yum.repos.d/目录下,ftp.repo 文件的内容如下:

```
[centos]
name=centos
gpgcheck=0
enabled=1
baseurl=ftp://192.168.200.16/centos
[gpmall]
name=gpmall
gpgcheck=0
enabled=1
baseurl=ftp://192.168.200.16/gpmall-single/gpmall-repo
```

可以看到该 repo 文件与 Redis 中的一样,不需要做其他配置。继续检查 Dockerfile 文件,复制完 YUM 源之后,安装数据库服务;复制 setup.sh 到容器内部,该脚本是数据库的操作脚本,setup.sh 的内容如下:

```
#!/bin/bash

mysqld_safe &
sleep 5
mysql -e "create database gpmall; use gpmall; source /root/gpmall.sql;"
mysql -e "grant all privileges on *.* to root@'localhost' identified by '123456';"
```

```
mysql -uroot -p123456 -e "grant all privileges on *.* to root@'%' identified
by '123456';"
```

该脚本做了启动数据库、创建库、导入库、设置权限等操作；Dockerfile 下一步操作为复制 gpmall.sql 文件到容器中，设置 setup.sh 的权限，运行脚本，最后启动 MySQL。

梳理完 Dockerfile 文件之后，进行镜像的制作，确认 /opt/mariadb 目录下存在 Dockerfile、ftp.repo、setup.sh 和 gpmall.sql 文件，且基本环境也配置完毕，制作镜像的命令如下：

```
[root@docker mariadb]# docker build -t gpmall-mariadb:v1.0 .
Sending build context to Docker daemon    64kB
Step 1/12 : FROM centos:centos7.5.1804
 ---> cf49811e3cdb
Step 2/12 : MAINTAINER chinaskill
 ---> Using cache
 ---> 778b2f97dc49
Step 3/12 : EXPOSE 3306
 ---> Running in 0ed78a0857c1
Removing intermediate container 0ed78a0857c1
 ---> 88c2ee09727c
...忽略输出...
Step 12/12 : CMD ["mysqld_safe"]
 ---> Running in 76ae0dd3ef91
Removing intermediate container 76ae0dd3ef91
 ---> 8a5e6739f80f
Successfully built 8a5e6739f80f
Successfully tagged gpmall-mariadb:v1.0
```

从以上 docker build 命令中可以看到制作的镜像名为 gpmall-mariadb:v1.0，没有报错，最后显示 Successfully built，说明镜像制作成功，可以查看镜像列表，如下所示：

```
[root@docker mariadb]# docker images
REPOSITORY           TAG       IMAGE ID        CREATED          SIZE
gpmall-mariadb       v1.0      8a5e6739f80f    2 minutes ago    833MB
```

制作完数据库镜像之后，接下来准备制作 ZooKeeper 服务的镜像。

（3）制作 ZooKeeper 镜像

想要制作 ZooKeeper 镜像，首先要创建工作目录，命令如下：

```
[root@docker ~]# mkdir /opt/zookeeper
```

根据手动安装 ZooKeeper 和配置的方式，编写 Dockerfile 文件，文件内容如下：

```
FROM centos:centos7.5.1804
MAINTAINER chinaskill
EXPOSE 2181
RUN rm -rf /etc/yum.repos.d/*
COPY ftp.repo /etc/yum.repos.d/ftp.repo
```

```
    RUN yum install java-1.8.0-openjdk java-1.8.0-openjdk-devel -y
    ADD zookeeper-3.4.14.tar.gz /opt
    RUN  mv  /opt/zookeeper-3.4.14/conf/zoo_sample.cfg  /opt/zookeeper-3.4.14/
conf/zoo.cfg
    CMD ["sh","-c","/opt/zookeeper-3.4.14/bin/zkServer.sh start && tail -f /etc/
shadow"]
```

从 Dockerfile 中可以看出，其使用了 CentOS7.5 的基础镜像；作者是 chinaskill；暴露了 2181 端口；删除容器内部原有的 repo 文件；复制 ftp.repo 到容器内部，ftp.repo 文件的内容如下：

```
[centos]
name=centos
gpgcheck=0
enabled=1
baseurl=ftp://192.168.200.16/centos
[gpmall]
name=gpmall
gpgcheck=0
enabled=1
baseurl=ftp://192.168.200.16/gpmall-single/gpmall-repo
```

ftp.repo 文件与之前无异，只要确认宿主机的 FTP 服务正常即可。回到 Dockerfile 文件，接下来安装 Java 环境，将 zookeeper-3.4.14.tar.gz 软件包传至容器内部的 /opt 目录下，将配置文件改名，最后运行 ZooKeeper 服务。

梳理完 Dockerfile 文件之后，进行镜像的制作，确认 /opt/zookeeper 目录下存在 Dockerfile、ftp.repo 和 zookeeper-3.4.14.tar.gz 文件，且基本环境也配置完毕，制作镜像的命令如下：

```
    [root@docker zookeeper]# docker build -t gpmall-zookeeper:v1.0 .
    Sending build context to Docker daemon  37.68MB
    Step 1/9 : FROM centos:centos7.5.1804
     ---> cf49811e3cdb
    Step 2/9 : MAINTAINER chinaskill
     ---> Using cache
     ---> 778b2f97dc49
    Step 3/9 : EXPOSE 2181
     ---> Running in 0ddf8537b8cc
    Removing intermediate container 0ddf8537b8cc
     ---> 8c7bdbad0b19
    ...忽略输出...
    Step 9/9 : CMD ["sh","-c","/opt/zookeeper-3.4.14/bin/zkServer.sh start &&
tail -f /etc/shadow"]
     ---> Running in b35e692223e4
    Removing intermediate container b35e692223e4
     ---> 3d2389d07944
    Successfully built 3d2389d07944
```

```
Successfully tagged gpmall-zookeeper:v1.0
```

同样地，从 docker build 命令中可以看到制作的镜像名为 gpmall-zookeeper:v1.0，没有报错，最后显示 Successfully built，说明镜像制作成功，可以查看镜像列表，如下所示：

```
REPOSITORY          TAG        IMAGE ID          CREATED           SIZE
gpmall-zookeeper    v1.0       3d2389d07944      2 minutes ago     540MB
```

制作完 ZooKeeper 镜像之后，接下来准备制作 Kafka 服务的镜像。

（4）制作 Kafka 镜像

制作 Kafka 镜像的方式有点特殊，因为 Kafka 服务需要依赖于 ZooKeeper 服务，也就是说，如果需要单独制作一个 Kafka 服务的镜像，则还需要在容器中提前安装 ZooKeeper 服务。其实 ZooKeeper 和 Kafka 这两个服务可以由一个镜像提供，感兴趣的读者可以尝试制作一个镜像，提供 ZooKeeper 和 Kafka 服务。

创建制作 Kafka 服务镜像的工作目录，命令如下：

```
[root@docker ~]# mkdir /opt/kafka
```

根据手动安装 Kafka 和配置的方式，编写 Dockerfile 文件，文件内容如下：

```
FROM centos:centos7.5.1804
MAINTAINER chinaskill
EXPOSE 9092
RUN rm -rf /etc/yum.repos.d/*
COPY ftp.repo /etc/yum.repos.d/ftp.repo
RUN yum install java-1.8.0-openjdk java-1.8.0-openjdk-devel -y
ADD zookeeper-3.4.14.tar.gz /opt
RUN mv /opt/zookeeper-3.4.14/conf/zoo_sample.cfg /opt/zookeeper-3.4.14/conf/zoo.cfg
ADD kafka_2.11-1.1.1.tgz /opt
CMD ["sh","-c","/opt/zookeeper-3.4.14/bin/zkServer.sh start && /opt/kafka_2.11-1.1.1/bin/kafka-server-start.sh /opt/kafka_2.11-1.1.1/config/server.properties"]
```

从以上 Dockerfile 中可以看出，它使用 CentOS7.5 基础镜像；作者是 chinaskill；暴露了 9092 端口；删除容器内部原有的 repo 文件；复制 ftp.repo 文件到容器内部；ftp.repo 文件的内容如下：

```
[centos]
name=centos
gpgcheck=0
enabled=1
baseurl=ftp://192.168.200.16/centos
[gpmall]
name=gpmall
gpgcheck=0
enabled=1
baseurl=ftp://192.168.200.16/gpmall-single/gpmall-repo
```

ftp.repo 文件仍然与之前的相同，回到 Dockerfile 文件，安装 Java 环境，将 zookeeper-3.4.14.tar.gz 软件包传至容器内部的 /opt 目录下，修改 ZooKeeper 的配置文件，再将 kafka_2.11-1.1.1.tgz 软件包传至容器内部的 /opt 目录下，启动 ZooKeeper 服务和 Kafka 服务。

　　梳理完 Dockerfile 文件之后，进行镜像的制作，确认 /opt/Kafka 目录下存在 Dockerfile、ftp.repo、zookeeper-3.4.14.tar.gz 和 kafka_2.11-1.1.1.tgz 文件，且基本环境也配置完毕，制作镜像的命令如下：

```
[root@docker kafka]# docker build -t gpmall-kafka:v1.0 .
Sending build context to Docker daemon  95.15MB
Step 1/10 : FROM centos:centos7.5.1804
 ---> cf49811e3cdb
Step 2/10 : MAINTAINER chinaskill
 ---> Using cache
 ---> 778b2f97dc49
Step 3/10 : EXPOSE 9092
 ---> Running in 6aa15cb815c1
...忽略输出...
Step 10/10 : CMD ["sh","-c","/opt/zookeeper-3.4.14/bin/zkServer.sh start && /opt/kafka_2.11-1.1.1/bin/kafka-server-start.sh /opt/kafka_2.11-1.1.1/config/server.properties"]
 ---> Running in 15cf48fb6b93
Removing intermediate container 15cf48fb6b93
 ---> 361c1f294c53
Successfully built 361c1f294c53
Successfully tagged gpmall-kafka:v1.0
```

　　同样地，从以上 docker build 命令中可以看到制作的镜像名为 gpmall-kafka:v1.0，没有报错，最后显示 Successfully built，说明镜像制作成功，可以查看镜像列表，如下所示：

```
[root@docker kafka]# docker images
REPOSITORY       TAG       IMAGE ID        CREATED         SIZE
gpmall-kafka     v1.0      361c1f294c53    4 minutes ago   603MB
```

　　制作完 Kafka 镜像之后，接下来准备制作 Nginx 服务的镜像。

（5）制作 Nginx 镜像

　　制作最后一个 Nginx 镜像，创建工作目录，命令如下：

```
[root@docker ~]# mkdir /opt/nginx
```

　　根据手动安装 Nginx 和配置的方式，编写 Dockerfile 文件，文件内容如下：

```
FROM centos:centos7.5.1804
MAINTAINER chinaskill
EXPOSE 80 443 8081 8082 8083
RUN rm -rf /etc/yum.repos.d/*
COPY ftp.repo /etc/yum.repos.d/ftp.repo
```

```
RUN yum install nginx java-1.8.0-openjdk java-1.8.0-openjdk-devel -y
COPY *.jar /root/
ADD dist.tar /root/
RUN rm -rf /usr/share/nginx/html/*
RUN mv /root/dist/* /usr/share/nginx/html/
COPY default.conf /etc/nginx/conf.d/default.conf
COPY setup.sh /root/setup.sh
RUN chmod 755 /root/setup.sh
CMD ["nginx","-g","daemon off;"]
```

从上述 Dockerfile 中可以看出,其使用了 CentOS7.5 的基础镜像;作者是 chinaskill;暴露了 80、443、8081、8082、8083 端口,其中 80 是 Web 服务端口,443 是 HTTPS 服务端口,8081/8083/8083 是后端服务的端口;删除容器内部原有的 repo 文件;复制 ftp.repo 文件到容器内部,ftp.repo 文件的内容如下:

```
[centos]
name=centos
gpgcheck=0
enabled=1
baseurl=ftp://192.168.200.16/centos
[gpmall]
name=gpmall
gpgcheck=0
enabled=1
baseurl=ftp://192.168.200.16/gpmall-single/gpmall-repo
```

ftp.repo 文件仍然与之前的相同。回到 Dockerfile 文件,安装 Java 环境,复制所有的 jar 包到容器内部的/root 目录下,将前端文件压缩包传至容器内部的/root 目录下,删除 Nginx 工作目录下原有的文件,将新的前端文件移动到 Nginx 工作目录,复制 Nginx 的 default.conf 替换容器中原有的配置文件,复制 setup.sh 到容器内,setup.sh 的内容如下所示:

```
#!/bin/bash

nohup java -jar /root/shopping-provider-0.0.1-SNAPSHOT.jar &
sleep 10
nohup java -jar /root/user-provider-0.0.1-SNAPSHOT.jar &
sleep 10
nohup java -jar /root/gpmall-shopping-0.0.1-SNAPSHOT.jar &
sleep 10
nohup java -jar /root/gpmall-user-0.0.1-SNAPSHOT.jar &
sleep 10
nginx
```

setup.sh 中的内容是启动 jar 包以及 Nginx。继续回到 Dockerfile 文件,对 setup.sh 赋予执行权限,最后在容器启动时,运行 Nginx 服务。

梳理完 Dockerfile 文件之后，进行镜像的制作，确认/opt/nginx 目录下存在 Dockerfile、ftp.repo、dist.tar、default.conf、setup.sh 和 4 个 jar 包文件，且基本环境也配置完毕，制作镜像命令如下：

```
[root@docker nginx]# docker build -t gpmall-nginx:v1.0 .
Sending build context to Docker daemon  215.6MB
Step 1/14 : FROM centos:centos7.5.1804
 ---> cf49811e3cdb
Step 2/14 : MAINTAINER chinaskill
 ---> Using cache
 ---> 778b2f97dc49
Step 3/14 : EXPOSE 80 443 8081 8082 8083
 ---> Running in 05690083082f
...忽略输出...
Step 14/14 : CMD ["nginx","-g","daemon off;"]
 ---> Running in 2ed6d0cb4d9d
Removing intermediate container 2ed6d0cb4d9d
 ---> a3b831dfcb12
Successfully built a3b831dfcb12
Successfully tagged gpmall-nginx:v1.0
```

从上面的 docker build 命令中可以看到制作的镜像名为 gpmall-nginx:v1.0，没有报错，最后显示 Successfully built，说明镜像制作成功，可以查看镜像列表，如下所示：

```
[root@docker nginx]# docker images
REPOSITORY      TAG     IMAGE ID        CREATED         SIZE
gpmall-nginx    v1.0    a3b831dfcb12    4 minutes ago   711MB
```

至此，所有的容器镜像制作完毕。接下来需要使用 docker-compose 编排工具，对这 5 个镜像进行编排，启动 gpmall 商城应用。

（6）编排部署 gpmall 商城

有了上述的 5 个容器镜像后，需要对这些镜像进行编排部署，首先创建 docker-compose 的工作目录，命令如下：

```
[root@docker opt]# mkdir gpmall
```

在/opt/gpmall 目录下，创建 docker-compose.yaml 文件，文件内容如下：

```
version: "3"
services:
  mysql.mall:
    container_name: mall-mysql
    image: gpmall-mariadb:v1.0
    ports:
      - "13306:3306"
    restart: always
```

```yaml
  redis.mall:
    container_name: mall-redis
    image: gpmall-redis:v1.0
    ports:
      - "16379:6379"
    restart: always

  zookeeper.mall:
    container_name: mall-zookeeper
    image: gpmall-zookeeper:v1.0
    ports:
      - "12181:2181"
    restart: always

  kafka.mall:
    depends_on:
      - zookeeper.mall
    container_name: mall-kafka
    image: gpmall-kafka:v1.0
    ports:
      - "19092:9092"
    restart: always

  mall:
    container_name: mall-nginx
    image: gpmall-nginx:v1.0
    links:
      - mysql.mall
      - redis.mall
      - zookeeper.mall
      - kafka.mall
    ports:
      - "83:80"
      - "1443:443"
    command: ["sh","-c","/root/setup.sh && tail -f /etc/shadow"]
```

从 Yaml 文件中可以看到，它启动了 5 个容器，前 4 个为基础服务容器，即 Redis、数据库、ZooKeeper 和 Kafka，分别指定了容器名、镜像名、映射端口和重启规则。最后启动的是 Nginx 容器，指定了镜像名和容器名，并依赖于前 4 个容器，在启动时执行容器内的 setup.sh 脚本。

编写完 docker-compose.yaml 文件后，进行编排启动，命令如下：

```
[root@docker gpmall]# docker-compose up -d
```

```
Creating network "gpmall_default" with the default driver
Creating mall-redis     ... done
Creating mall-zookeeper ... done
Creating mall-mysql     ... done
Creating mall-kafka     ... done
Creating mall-nginx     ... done
```

docker-compose 编排启动成功后，可以看到容器正常启动，查看端口开放情况，命令如下：

```
[root@docker gpmall]# netstat -ntpl
Active Internet connections (only servers)
Proto Recv-Q Send-Q Local Address      Foreign Address    State       PID/Program name
tcp       0      0 0.0.0.0: 22         0.0.0.0:*          LISTEN      1025/sshd
tcp       0      0 127.0.0.1:25        0.0.0.0:*          LISTEN      1292/master
tcp6      0      0 : : :1443           :::*               LISTEN      15992/docker-proxy
tcp6      0      0 : : :9092           :::*               LISTEN      6164/java
tcp6      0      0 : : :39141          :::*               LISTEN      6164/java
tcp6      0      0 : : :2181           :::*               LISTEN      5856/java
tcp6      0      0 : : :83             :::*               LISTEN      160037docker-proxy
tcp6      0      0 : : :19092          :::*               LISTEN      15832/docker-proxy
tcp6      0      0 : : :12181          :::*               LISTEN      15604/docker-proxy
tcp6      0      0 : : :21             :::*               LISTEN      6440/vsftpd
tcp6      0      0 : : :22             :::*               LISTEN      1025/sshd
tcp6      0      0 : : :25             :::*               LISTEN      1292/master
tcp6      0      0 : : :13306          :::*               LISTEN      15582/docker-proxy
tcp6      0      0 : : :16379          :::*               LISTEN      15569/docker-proxy
tcp6      0      0 : : :33211          :::*               LISTEN      5856/java
```

端口开放情况正常，可以使用浏览器访问 http://192.168.200.16:83，如图 6-3 所示。

图 6-3　商城首页

看到图 6-3 则说明使用 docker-compose 编排工具启动 gpmall 应用商城成功。如果访问页面失败，一般来说是因为 Nginx 容器中的 jar 包没有正常启动（机器配置有区别，启动 jar

间隙太短可能导致 jar 包没有正常启动），可以使用 docker exec 命令进入 Nginx 容器，查看 jar 包是否都正常启动，若没有正常启动，则可以使用手动的方式启动 jar 包，或者修改 setup.sh 中的 sleep 时间。

归纳总结

通过本单元的教学内容，读者完整地学习了 Docker Compose 的案例，从最开始的手动部署商城应用，到制作容器镜像、编排部署，可以体验到使用容器与使用虚拟机的不同。使用容器编排部署，效率更高且不容易出错，一旦准备好容器镜像，只需要编写编排文件即可进行部署。

课后练习

一、判断题

1. 在 docker-compose 配置文件中指定 Compose 文件的版本的参数是 version。（　　）
2. 在 docker-compose 配置文件中指定 Dockerfile 文件所在的路径的参数是 context。（　　）

二、单项选择题

1. 在 docker-compose 配置文件中定义尝试重启的间隔时间的参数是（　　）。
   A. name　　　　　　B. container_name　　C. volumes　　　　D. delay
2. 在 docker-compose 配置文件中指定设备映射列表的参数是（　　）。
   A. name　　　　　　B. container_name　　C. devices　　　　D. delay

三、多项选择题

1. 下面关于容器编排的说法，不正确的是（　　）。
   A. 容器编排是指对单独组件和应用层的工作进行组织的流程
   B. 应用一般由单独容器化的组件（通常称为微服务）组成
   C. 对单个容器进行组织的流程即称为容器编排
   D. 容器编排工具仅允许用户指导容器部署与自动更新
2. 下面关于 Docker Compose 的说法中，不正确的是（　　）。
   A. Docker Compose 是对 Docker 容器进行编排的工具
   B. Docker Compose 通过多个配置文件来管理多个 Docker 容器
   C. Docker Compose 默认的模板文件是 docker-compose.py
   D. Docker Compose 常用命令可以使用 docker-compose -help 命令进行查询帮助

技能训练

1. 成功部署 docker-compose，并使用命令查询其版本。
2. 通过编排服务部署 gpmall 商城服务，并使用 curl 命令访问其主页。

单元 7　Docker Swarm 编排服务

学习目标

通过本单元的学习，要求了解 Docker Swarm 容器集群服务的基本概念、架构与特点；Docker Swarm 的使用方法与技巧等。本单元培养读者掌握 Docker Swarm 集群服务在工作中的实际运用、掌握 Docker Swarm 集群的搭建使用和运维实战等技能，也培养读者的语言学习和信息检索的能力。

7.1　认识 Docker Swarm 编排服务

1. Docker Swarm 简介

Docker 容器编排（1）

Swarm 是 Docker 公司推出的用来管理 Docker 集群的平台，几乎全部用 Go 语言来完成开发，它将一群 Docker 宿主机变成一个单一的虚拟主机。Swarm 使用标准的 Docker API 接口作为其前端的访问入口，换言之，各种形式的 Docker Client（compose、docker-py 等）均可以直接与 Swarm 通信，甚至 Docker 本身都可以很容易地与 Swarm 集成，这大大方便了用户将原本基于单节点的系统移植到 Swarm 上。同时 Swarm 内置了对 Docker 网络插件的支持，用户也可以很容易地部署跨主机的容器集群服务。

Docker Swarm 和 Docker Compose 一样，都是 Docker 官方容器编排项目，但不同的是，Docker Compose 是一个在单个服务器或主机上创建多个容器的工具，而 Docker Swarm 则可以在多个服务器或主机上创建容器集群服务，对于微服务的部署，显然 Docker Swarm 会更加适合。

从 Docker 1.12.0 版本开始，Docker Swarm 包含在 Docker 引擎中，并且已经内置了服务发现工具，用户就不需要像之前一样，再配置 Etcd 或者 Consul 来进行服务发现配置了。

Swarm Deamon 只是一个调度器（Scheduler）加路由器（Router），Swarm 自己不运行容器，它只是接收 Docker 客户端发来的请求，调度适合的节点来运行容器。这就意味着，即使 Swarm 由于某些原因崩溃了，集群中的节点也会照常运行，Swarm 重新恢复运行之后会收集重建集群信息。

2. Docker Swarm 架构

Docker Swarm 的架构图如图 7-1 所示。

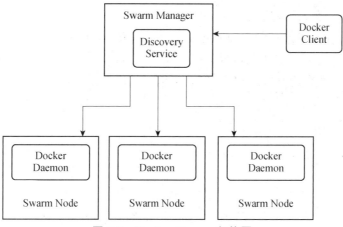

图 7-1 Docker Swarm 架构图

在架构图中可以看出 Docker Client 使用 Swarm 对集群（Cluster）进行调度使用。从图 7-1 还可以看出 Swarm 是典型的 master-slave 结构,通过发现服务来选举 Swarm Manager。Swarm Manager 是中心管理节点,各个 Swarm Node 上运行 Agent 接受 Manager 的统一管理。

Docker Swarm 的架构解析如图 7-2 所示。

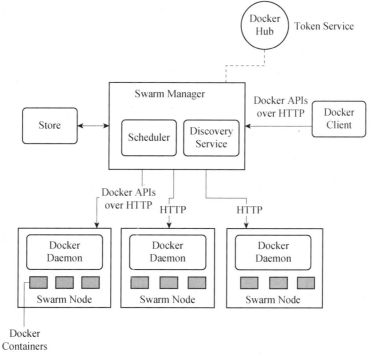

图 7-2 Docker Swarm 架构解析

图 7-2 所示 Docker Swarm 作为一个整体实际上处于一个集群中,它可能对应了一到多台的服务器。每台服务器上都装有 Docker 并且开启了基于 HTTP 的 Docker API。这个集群中有一个 Swarm Manager 管理节点,用来管理集群中的容器资源。管理者的管理对象不是服务器层面而是集群层面的,也就是说,通过 Swarm Manager,只能笼统地向集群发出指

令，而不能具体到某台具体的服务器需要干什么。

Docker Swarm 具体的管理实现方式是 Swarm Manager 节点向外暴露了一个 HTTP 接口，外部用户通过这个 HTTP 接口来实现对集群的管理。对于稍微大一点的集群，最好使用一台单独的服务器作为专门的管理节点，对于学习而言，也可以把管理节点和工作节点放在一台服务器上。

3. Docker Swarm 的基本概念

Docker Swarm 的基本概念介绍如下。

（1）节点

运行 Docker 的主机可以主动初始化一个 Swarm 集群或者加入一个已存在的 Swarm 集群，这台运行 Docker 的主机就成为一个 Swarm 集群的节点（Node）。

节点分为管理（Manager）节点和工作（Worker）节点。

管理节点用于 Swarm 集群的管理，Docker Swarm 的命令基本只能在管理节点执行（节点退出集群命令 docker swarm leave 可以在工作节点执行）。一个 Swarm 集群可以有多个管理节点，但只有一个管理节点可以成为 Leader，Leader 通过 raft 协议实现。

工作节点是任务执行节点，管理节点将服务（Service）下发至工作节点执行。管理节点默认作为工作节点。

节点的工作模式如图 7-3 所示。

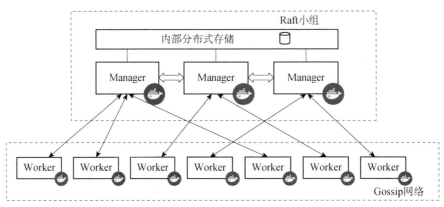

图 7-3　Docker Swarm 工作模式

（2）服务和任务

任务（Task）是 Swarm 中的最小调度单位，目前来说就是一个单一的容器。

服务（Services）是指一组任务的集合，服务定义了任务的属性。服务有两种模式，分别如下。

① replicated services：按照一定规则在各个工作节点上运行指定个数的任务。

② global services：每个工作节点上运行一个任务。

两种模式通过 docker service create 命令中的 --mode 参数指定。

服务与任务的工作模式，如图 7-4 所示。

通过上述对 Docker Swarm 的介绍，读者对 Swarm 的产生、节点的作用、基本概念有了一定的认识，接下来使用虚拟机搭建 Swarm 集群，体验 Swarm 集群的作用。

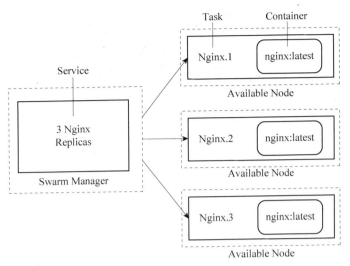

图 7-4 服务与任务的工作模式示意图

7.2 使用 Docker Swarm 编排服务

1. 环境准备

（1）规划节点

Docker Swarm 实操案例节点规划如表 7-1 所示。

表 7-1 Docker Swarm 实操案例节点规划

IP 地址	主机名	节点
192.168.200.17	master	Swarm 管理节点
192.168.200.18	node	Swarm 工作节点

（2）基础准备

准备好 VMware Workstation 工具和 CentOS7.5 版本虚拟机。使用 CentOS7.5 虚拟机快照，克隆两台虚拟机，云主机配置自定义，推荐配置 2vCPU、4GB 内存、40GB 硬盘，网络使用 NAT 模式，master 节点 IP 设置为 192.168.200.17，node 节点 IP 设置为 192.168.200.18。推荐设置主机密码为 000000。注意，请自行将虚拟机配置好 IP、安装 Docker 服务，关闭 SELinux 和防火墙服务。

2. Docker Swarm 集群搭建

（1）基础准备

根据案例分析中的节点规划准备两台虚拟机，使用远程连接工具连接到两台虚拟机，并分别修改两个节点主机名为 master 和 node，修改完之后如下所示：

master 节点：

```
[root@master ~]# hostnamectl
   Static hostname: master
        Icon name: computer-vm
```

```
           Chassis: vm
        Machine ID: 7378af07fbd948cf8fea45ebcaff9ada
           Boot ID: 861826e76c694be0952502c95fa3a3dc
    Virtualization: vmware
  Operating System: CentOS Linux 7 (Core)
       CPE OS Name: cpe:/o:centos:centos:7
            Kernel: Linux 3.10.0-862.el7.x86_64
      Architecture: x86-64
```

node 节点：

```
[root@node ~]# hostnamectl
   Static hostname: node
         Icon name: computer-vm
           Chassis: vm
        Machine ID: 7378af07fbd948cf8fea45ebcaff9ada
           Boot ID: 38921aca4e96405e80411611889a267e
    Virtualization: vmware
  Operating System: CentOS Linux 7 (Core)
       CPE OS Name: cpe:/o:centos:centos:7
            Kernel: Linux 3.10.0-862.el7.x86_64
      Architecture: x86-64
```

修改完主机名之后，修改两个节点的/etc/hosts 文件，如下所示：

```
# cat /etc/hosts
127.0.0.1    localhost localhost.localdomain localhost4 localhost4.localdomain4
::1          localhost localhost.localdomain localhost6 localhost6.localdomain6
#在/etc/hosts 文件中添加如下两行
192.168.200.17  master
192.168.200.18  node
```

（2）安装 Docker

两个节点都安装 Docker 服务，安装 Docker 的操作在前面的单元中已叙述，此处不再赘述，安装完之后，如下所示：

master 节点：

```
[root@master ~]# docker --version
Docker version 19.03.13, build 4484c46d9d
```

node 节点：

```
[root@node ~]# docker --version
Docker version 19.03.13, build 4484c46d9d
```

（3）配置 Docker API

master 节点和 node 节点均需开启 Docker API，命令如下：

```
# vi /lib/systemd/system/docker.service
```
将 fd:// --containerd=/run/containerd/containerd.sock
修改为 tcp://0.0.0.0:2375 -H unix:///var/run/docker.sock
```
# systemctl daemon-reload
# systemctl restart docker
```

两个节点均执行上述操作。

（4）部署 Swarm 集群

在 master 节点创建 Swarm 集群，命令如下：

```
[root@master ~]# docker swarm init --advertise-addr 192.168.200.17
Swarm initialized: current node (v5zrzopkzv3eeytm433s8v7q5) is now a manager.

To add a worker to this swarm, run the following command:

    docker swarm join --token SWMTKN-1-1klop5tk8v4972d21oipbrgazg6roslq1f
cuubtpjtvbkrt7u1-eb4cay4flwcfix5u3yc2xw0va 192.168.200.17:2377

To add a manager to this swarm, run 'docker swarm join-token manager' and
follow the instructions.
```

可以看到，安装 Docker Swarm 时不需要安装其他的软件包，因为 Swarm 服务已经集成在 Docker 中了。从上面的反馈结果来看，此时 master 节点已经被初始化成 Manager 管理节点了，初始化命令中"--advertise-addr"表示管理节点的 IP。其他节点必须能通过这个 IP 找到管理节点，输出结果中包含了加入 Swarm 集群的命令。接下来就是将 node 节点加入集群中，在 node 节点输入如下命令：

```
[root@node ~]# docker swarm join --token SWMTKN-1-1klop5tk8v4972d21oip
brgazg6roslq1fcuubtpjtvbkrt7u1-eb4cay4flwcfix5u3yc2xw0va 192.168.200.17:2377
This node joined a swarm as a worker.
```

node 节点成功作为一个工作节点加入 Swarm 集群。

在 master 节点，查看 Swarm 集群中各个节点的状态，如下所示：

```
[root@master -]# docker node update --availability drain node
[root@master ~]#docker node ls
ID                          HOSTNAME   STATUS   AVAILABILITY   MANAGER STATUS   ENGINE VERSION
5zrzopkzv3eeytm433s8v7q5*   master     Ready    Active         Leader           19.03.13
r61yolrwnxccj25pfge2dedxz   node       Ready    Active                          19.03.13
```

可以看到 master 和 node 节点的状态都是 Active。节点的 AVAILABILITY 状态是可以更改的，除了 Active 状态外，还有 Drain 状态，这两个状态的说明如下：

● Active 状态下，node 可以接收来自 manager 节点的任务分派。

● Drain 状态下，node 节点会结束 Task（任务），且不再接收来自 Manager 节点的任务分派（也就是下线节点）。

将 node 节点修改为下线节点，命令如下：

```
[root@master -]# docker node update --availability drain node
[root@master ~]#docker node ls
ID                          HOSTNAME   STATUS   AVAILABILITY   MANAGER STATUS   ENGINE VERSION
5zrzopkzv3eeytm433s8v7q5*   master     Ready    Active         Leader           19.03.13
r61yolrwnxccj25pfge2dedxz   node       Ready    Drain                           19.03.13
```

如上所示，当 node 的状态改为 Drain 后，该节点就不会接收 Task 任务分发，就算之前已经接收了任务，也会转移到别的节点上。再次将其修改为 Active 状态及将下线的节点再次上线，命令如下：

```
[root@master -]# docker node update --availability drain node
[root@master ~]#docker node ls
ID                          HOSTNAME   STATUS   AVAILABILITY   MANAGER STATUS   ENGINE VERSION
5zrzopkzv3eeytm433s8v7q5*   master     Ready    Active         Leader           19.03.13
r61yolrwnxccj25pfge2dedxz   node       Ready    Active                          19.03.13
```

至此，Docker Swarm 集群安装完毕。下面介绍 Docker Swarm 的图形化管理工具 Portainer。

(5) 部署图形化管理工具

Portainer 是 Docker 的图形化管理工具，提供状态显示面板、应用模板快速部署、容器镜像网络数据卷的基本操作（包括上传下载镜像、创建容器等操作）、事件日志显示、容器控制台操作、Swarm 集群和服务等集中管理和操作、登录用户管理和控制等功能。它的功能十分全面，基本能满足中小型企业对容器管理的全部需求。

① 创建存储卷。在 master 节点和 node 节点创建 Portainer 后端存储卷，命令如下：

```
[root@master ~]# docker volume create portainer_data
portainer_data
[root@node ~]# docker volume create portainer_data
portainer_data
```

② 上传镜像。在 master 节点和 node 节点分别加载 Portainer 镜像，命令如下：

```
[root@master ~]# docker load -i portainer_portainer-ce_latest.tar
[root@master ~]# docker load -i portainer_agent_latest.tar
[root@node ~]# docker load -i portainer_agent_latest.tar
```

在 master 节点和 node 节点分别查看镜像列表，命令如下：

master 节点：

```
[root@master ~]# docker images
REPOSITORY                  TAG      IMAGE ID        CREATED      SIZE
portainer/portainer-ce      latest   8377e6877145    4 days ago   251MB
portainer/agent             latest   861177a970d1   4 days ago   156MB
```

node 节点：

```
[root@node ~]# docker images
```

```
REPOSITORY        TAG       IMAGE ID       CREATED       SIZE
portainer/agent   latest    861177a970d1   4 days ago    156MB
```

③ 编写模板文件。在 master 节点编写模板文件 portainer-stack.yaml，创建命令与文件内容如下：

```
[root@master ~]# vi portainer-stack.yaml
[root@master ~]# cat portainer-stack.yaml
version: '3.2'
services:
  agent:
    image: portainer/agent
    volumes:
      - /var/run/docker.sock:/var/run/docker.sock
      - /var/lib/docker/volumes:/var/lib/docker/volumes
    networks:
      - agent_network
    deploy:
      mode: global
      placement:
        constraints: [node.platform.os == linux]
  portainer:
    image: portainer/portainer-ce
    command: -H tcp://tasks.agent:9001 --tlsskipverify
    ports:
      - "9000:9000"
      - "8000:8000"
    volumes:
      - portainer_data:/data
    networks:
      - agent_network
    deploy:
      mode: replicated
      replicas: 1
      placement:
        constraints: [node.role == manager]
networks:
  agent_network:
    driver: overlay
    attachable: true
volumes:
  portainer_data:
```

编辑完 Yaml 文件后，通过模板文件创建 Portainer 服务，命令如下：

```
[root@master ~]# docker stack deploy -c portainer-stack.yaml portainer
```

```
Creating network portainer_agent_network
Creating service portainer_agent
Creating service portainer_portainer
```

服务创建完毕后，通过浏览器访问 Portainer 服务，在浏览器上输入 http://192.168.200.17:9000，访问 Portainer 服务界面，如图 7-5 所示。

图 7-5　Portainer 服务界面

在该界面填写相关信息，设置用户名和密码，再输入确认密码，然后单击 "Create user" 按钮，即可进入控制台，如图 7-6 所示。

图 7-6　控制台

至此，Portainer 图形化界面服务安装完毕，在平时工作中使用 Docker Swarm 集群的时候，一般通过命令行完成，也可以通过 Portainer 界面操作，关于界面操作的具体步骤，此处不再赘述，感兴趣的读者可以自行学习。下面通过一些操作命令和实操案例，让读者更深入地了解 Docker Swarm 集群。

3. Docker Swarm 集群的使用

（1）Swarm 部署服务

使用 Docker Swarm 部署服务，此处以部署 Nginx 为例，先将提供的 nginx_latest.tar 镜像上传至 master 节点的/root 目录下，然后上传，命令如下：

Docker 容器编排（2）

```
[root@master ~]# docker load -i nginx_latest.tar
07cab4339852: Loading layer [==================>]  72.49MB/72.49MB
822ae9fef1d8: Loading layer [==================>]  64.53MB/64.53MB
7230cfe05cc1: Loading layer [==================>]  3.072kB/3.072kB
8eb80f066de2: Loading layer [==================>]  4.096kB/4.096kB
8032102adebe: Loading layer [==================>]  3.584kB/3.584kB
Loaded image: nginx:latest
```

创建 Nginx 使用的网络，命令如下：

```
[root@master opt]# docker network create -d overlay nginx_net
jgm35litfmdtjrge074jtr18n
```

查看容器网络，命令如下：

```
[root@master opt]# docker network ls |grep nginx_net
jgm35litfmdt        nginx_net           overlay             swarm
```

部署 Nginx 服务，设置副本数量为 1，使用刚才创建的网络，命令如下：

```
[root@master ~]# docker service create --replicas 1 --network nginx_net --name my_nginx -p 80:80 nginx
7vy9txe1gjemy6qkwwkfbkjzt
overall progress: 1 out of 1 tasks
1/1: running   [==========================================>]
verify: Service converged
```

部署完成后，查看正在运行服务的列表，命令如下：

```
[root@master ~]# docker service ls
ID              NAME       MODE         REPLICAS   IMAGE           PORTS
7vy9txe1gjem    my_nginx   replicated   1/1        nginx:latest    *:80->80/tcp
```

可以看到 Nginx 应用已经启动，查询 Swarm 中服务的信息，命令如下：

```
[root@master ~]# docker service inspect --pretty my_nginx
ID:             7vy9txe1gjemy6qkwwkfbkjzt
Name:           my_nginx
Service Mode:   Replicated
 Replicas:      1
Placement:
UpdateConfig:
 Parallelism:   1
```

```
  On failure: pause
  Monitoring Period: 5s
  Max failure ratio: 0
  Update order:      stop-first
 RollbackConfig:
  Parallelism:    1
  On failure: pause
  Monitoring Period: 5s
  Max failure ratio: 0
  Rollback order:    stop-first
 ContainerSpec:
  Image:
  nginx:latest@sha256:853b221d3341add7aaadf5f81dd088ea943ab9c918766e295321294b035f3f3e
  Init:       false
 Resources:
 Networks: nginx_net
 Endpoint Mode:   vip
 Ports:
  PublishedPort = 80
  Protocol = tcp
  TargetPort = 80
  PublishMode = ingress
```

使用--pretty 参数使输出格式化为可读的格式，不加—pretty 参数则可以输出更详细的信息，通过该命令可以看到服务的详细信息。

还可以使用 ps 命令，查看服务被部署到哪个节点了，命令如下：

```
[root@master ~]# docker service ps my_nginx
ID    NAME    IMAGE    NODE    DESIRED STATE    CURRENT STATE    ERROR    PORTS
vqz76bymghul  my_nginx.1  nginx:latest  node  Running  Running 2 hours ago
```

可以看到 Nginx 服务被部署到了 node 节点，通过 node 节点的 IP 可以访问 Nginx 应用，用浏览器访问 http://192.168.200.18，如图 7-7 所示。

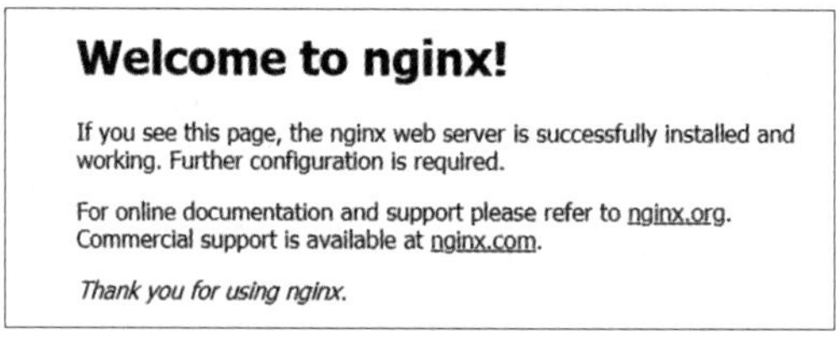

图 7-7 Nginx 首页

（2）Swarm 动态扩容服务

当然，如果只是通过 Service 启动容器，Swarm 也算不上什么新鲜的东西了。Service 还提供了复制（类似 Kubernetes 里的副本）功能，可以通过 docker service scale 命令来设置服务中容器的副本数，比如将上面的 my_nginx 容器动态扩展到 4 个，命令如下：

```
[root@master ~]# docker service scale my_nginx=4
my_nginx scaled to 4
overall progress: 4 out of 4 tasks
1/4: running   [==================================>]
2/4: running   [==================================>]
3/4: running   [==================================>]
4/4: running   [==================================>]
verify: Service converged
```

和创建服务一样，增加 scale 数之后，将会创建新的容器，这些新启动的容器也会经历从准备到运行的过程，过一小段时间之后，服务应该就会启动完成，这时候可以再来看一下容器数量，命令如下：

```
[root@master ~]# docker service ps my_nginx
ID              NAME            IMAGE           NODE    DESIRED STATE  CURRENT STATE
vqz76bymghulmy_nginx.1          nginx:latest    node    Running        Running 2 hours ago
lbcnreuh3niemy_nginx.2          nginx:latest    master  Running        Running 2 minutes ago
eo0klotnbkkumy_nginx.3          nginx:latest    master  Running        Running 2 minutes ago
qw988xfur6tfmy_nginx.4          nginx:latest    node    Running        Running 3 minutes ago
```

可以看到，之前 my_nginx 容器只在 node 节点上有 1 个实例，而现在又增加了 3 个实例。这 4 个副本的 my_nginx 容器分别运行在这两个节点上，登录这两个节点，就会发现已经存在运行着的 my_nginx 容器。

（3）模拟 node 节点宕机

在 Docker Swarm 集群中，如果一个节点宕机了（即该节点从 Swarm 集群中被踢出），则会将在该节点运行的容器调度到其他节点，以满足指定数量的副本保持运行状态。

比如将 node 节点宕机或将 node 节点上的 Docker 服务关闭，那么它上面的 Task 实例就会转移到别的节点上。当 node 节点恢复后，它转移出去的 Task 实例不会主动转移回来，只能等别的节点出现故障后再转移 Task 实例到它的上面。

下面将模拟 node 节点宕机，先将 node 节点上的 Docker 服务关闭，命令如下：

```
[root@node ~]# systemctl stop docker
```

然后在 master 节点查看 Swarm 集群中的节点，命令如下：

```
[root@master ~]#docker node ls
ID                            HOSTNAME  STATUS  AVAILABILITY  MANAGER STATUS  ENGINE VERSION
5zrzopkzv3eeytm433s8v7q5*     master    Ready   Active        Leader          19.03.13
r61yolrwnxccj25pfge2dedxz     node      down    Active                        19.03.13
```

可以看到 node 节点此时的状态变为了 down。接着查询 my_nginx 下的容器，命令如下：

```
[rootdmaster    ~]#docker service ps my_nginx
    ID              NAME            IMAGE           NODE        DESIRED STATE   CURRENT STATE
    5bisgi5otayj    my_nginx.1      nginx:latest    master      Running         Running 2
minutes ago
    vqz76bymghul    my_nginx.1      nginx:latest    node        Shutdown        Running 2
hours ago
    lbcnreuh3nie    my_nginx.2      nginx:latest    master      Running         Running 18
minutes ago
    eo0klotnbkku    my_nginx.3      nginx:latest    master      Running         Running 18
minutes ago
    5zjpraecp6ca    my_nginx.4      nginx:latest    master      Running         Running 2
minutes ago
    qw988xfur6tf    my_nginx.4      nginx:latest    node        Shutdown        Running 19
minutes ago
```

可以看到原来运行在 node 节点上的容器状态变成了 Shutdown（关闭），且容器都转移到 master 节点上运行了。

回到 node 节点，先将 node 节点的 Docker 服务启动，命令如下：

```
[root@node ~]# systemctl start docker
```

然后查看 Swarm 集群中的节点状态，命令如下：

```
[root@master ~]#docker node ls
ID                          HOSTNAME    STATUS  AVAILABILITY    MANAGER STATUS  ENGINE VERSION
5zrzopkzv3eeytm433s8v7q5*   master      Ready   Active          Leader          19.03.13
r61yolrwnxccj25pfge2dedxz   node        Ready   Active                          19.03.13
```

可以看到 node 节点的状态又变回了 Active 状态，切换到 master 节点，查看运行的容器，命令如下：

```
[root@master ~]# docker ps -a
    CONTAINER ID        IMAGE               COMMAND
CREATED             STATUS              PORTS               NAMES
    6a14d6ed7e93        nginx:latest        "/docker-entrypoint.…"
12 minutes ago      Up 12 minutes       80/tcp              my_nginx.1.
5bisgi5otayjyfu4supsk316i
    8ece5031e3dd        nginx:latest        "/docker-entrypoint.…"
12 minutes ago      Up 12 minutes       80/tcp              my_nginx.4.
5zjpraecp6ca5737t1gm8damt
    6d0cd2450769        nginx:latest        "/docker-entrypoint.…"
28 minutes ago      Up 28 minutes       80/tcp              my_nginx.3.
eo0klotnbkku8a0gdqpublykd
    591d7b11d4d9        nginx:latest        "/docker-entrypoint.…"
28 minutes ago      Up 28 minutes       80/tcp              my_nginx.2.
lbcnr0uh3niebgs2e8l88ysxo
```

将原来运行在 node 节点上的 my_nginx.1 和 my_nginx.4 关闭，查看这两个容器是否会迁移回 node 节点，命令如下：

```
[root@master ~]# docker stop 6a14d6ed7e93 8ece5031e3dd
6a14d6ed7e93
8ece5031e3dd
```

关闭这两个容器后，查看容器是否回到了 node 节点，命令如下：

```
[rootdmaster    ~]#docker service ps my_nginx
ID              NAME             IMAGE         NODE    DESIRED STATE   CURRENT STATE
vtfplwlsmyvx    my_nginx.1    nginx:latest  node  Running       Running 2 minutes ago
5bisgi5otayj    my_nginx.1    nginx:latest  master Shutdown complete about a minute ago
vqz76bymghul    my_nginx.1  nginx:latest  node   Shutdown      Shutdown 9 minutes ago
lbcnreuh3nie    my_nginx.2  nginx:latest    master  Running Running 18 minutes ago
eo0klotnbkku    my_nginx.3  nginx:latest    master  Running Running 18 minutes ago
5zjpraecp6ca    my_nginx.4  nginx:latest    master Shutdown complete about a minute ago
qw988xfur6tf    my_nginx.4  nginx:latest    node    Shutdown  Shutdown  9 minutes ago
u1uz3fej2443    my_nginx.4  nginx:latest    node  Running    Running 18 minutes ago
```

可以发现这两个容器又回到 node 节点运行了。从这个案例可以看出，在 Swarm cluster 集群中启动的容器，在工作节点被删除或停用后，该容器会自动转移到其他的工作节点上。

（4）Swarm 动态缩容服务

在上面的案例中，使用了 Swarm 动态扩容的功能，同理 Swarm 还可以缩容，使用的同样是 scale 命令。将 my_nginx 的容器数量缩小至 1 个，命令如下：

```
[root@master ~]# docker service scale my_nginx=1
my_nginx scaled to 1
overall progress: 1 out of 1 tasks
1/1: running   [==================================>]
verify: Service converged
```

查看 my_nginx 中的容器数量，命令如下：

```
[root@master ~]# docker service ps my_nginx
ID              NAME          IMAGE         NODE         DESIRED STATE  CURRENT STATE
vtfplwlsmyvx    my_nginx.1    nginx:latest  node         Running
```

```
Running            27 minutes ago
    5bisgi5otayj       \_ my_nginx.1     nginx:latest    master       Shutdown
Complete           27 minutes ago
    vqz76bymghul       \_ my_nginx.1     nginx:latest    node         Shutdown
Shutdown           35 minutes ago
    5zjpraecp6ca          my_nginx.4     nginx:latest    master       Shutdown
Complete           27 minutes ago
    qw988xfur6tf       \_ my_nginx.4     nginx:latest    node         Shutdown
Shutdown           35 minutes ago
```

可以看到系统中只剩下一个 my_nginx.1 在运行，其他容器都停止了。登录 node 节点查看容器的状态，命令如下：

```
[root@node ~]# docker ps -a
CONTAINER ID      IMAGE                COMMAND                  CREATED
STATUS                PORTS            NAMES
    5c25ca77248e      nginx:latest         "/docker-entrypoint.…"   29
minutes ago      Up 29 minutes           80/tcp           my_nginx.1.
vtfplwlsmyvx9e7spdjr2gb3g
    93a5d98ef362      nginx:latest         "/docker-entrypoint.…"   About
an hour ago      Exited (0) 44 minutes ago                my_nginx.4.
qw988xfur6tfgh4qvl87atl5k
    9142b1ecf04b      nginx:latest         "/docker-entrypoint.…"   3 hours
ago              Exited (0) 44 minutes ago                my_nginx.1.vqz76
bymghulki1k5pylxu27f
```

查看容器状态，可以发现容器停止了，而不是被删除了。

（5）Swarm 的 update 命令

在上面的案例中，介绍了使用 scale 命令对服务进行扩容或者缩容。除了使用 scale 进行容器的扩容或缩容之外，还可以使用 docker service update 命令，对服务的启动参数进行更新/修改。假设现在要将 my_nginx 中的容器副本数量扩容至 3，可以使用如下命令：

```
[root@master ~]# docker service update --replicas 3 my_nginx
my_nginx
overall progress: 3 out of 3 tasks
1/3: running   [==================================================>]
2/3: running   [==================================================>]
3/3: running   [==================================================>]
verify: Service converged
```

扩容之后，查看 my_nginx 服务下的容器数量，命令如下：

```
[[rootdmaster   ~]#docker service ps my_nginx
ID              NAME           IMAGE           NODE      DESIRED STATE   CURRENT STATE
    vtfplwlsmyvx    my_nginx.1     nginx:latest    node      Running         Running 51
minutes ago
```

```
    5bisgi5otayj       1my_nginx.1  nginx:latest    master   Shutdown    complete
51 minutes ago
    vqz76bymghui       my_nginx.1   nginx:latest    node     Shutdown    Shutdown
59 minutes ago
    u6nx5qldoueg       my_nginx.2   nginx:latest    master   Running     Running
about a minute ago
    nzk64vvq7gex       my_nginx.3   nginx: latest   node     Running     Running
about a minute ago
```

可以看到现在有 3 个正在运行的容器。docker service update 命令也可用于直接升级镜像等操作，假如当前存在 nginx:new 版本的镜像，想将原来的 nginx:latest 替换掉，可以使用如下命令：

```
[root@master ~]# docker service update --image nginx: new my_nginx
```

（6）删除 Swarm 服务

如果要将部署的 Nginx 应用删除，可以使用如下命令：

```
[root@master ~]# docker service rm my_nginx
my_nginx
```

删除完之后，会把所有节点上的所有容器（task 任务实例）全部删除。

（7）Swarm 的负载均衡功能

测试 Swarm 集群的负载均衡功能，继续使用 Nginx 镜像进行实验。在 Swarm 集群中，创建一个 Web 服务，使用 Nginx 镜像，副本数为 2，命令如下：

```
[root@master ~]# docker service create --name web --replicas 2 -p 80:80 nginx:latest
dez6o0d96clpr3sipzkm2sdtr
overall progress: 2 out of 2 tasks
1/2: running   [==================================>]
2/2: running   [==================================>]
verify: Service converged
```

创建完成后，查看 Web 服务中的容器，命令如下：

```
[root@master -]# docker service ps web
ID             NAME        IMAGE          NODE    DESIRED STATE  CURRENT STATE
kprafwrstlfu   web.1       nginx:latest   master  Running        Running about a
minute ago
    p4f5p470vjqx   web.2   nginx:latest   node    Running        Running about a
minute ago
```

可以看到一个容器在 master 节点启动，一个容器在 node 节点启动。在 master 节点创建 index.html 文件，文件内容如下：

```
[root@master ~]# vi index.html
[root@master ~]# cat index.html
```

web1.server

将该文件上传至容器内部的 nginx 工作目录，命令如下：

```
[root@master ~]# docker container cp index.html c9a8c795fb51:/usr/share/nginx/html/
```

容器的 ID 可以通过 docker ps –a 命令进行查看。

切换到 node 节点，同样创建 index.html 文件，文件内容如下：

```
[root@node ~]# vi index.html
[root@node ~]# cat index.html
web2.server
```

将 index.html 文件复制到 node 节点的 Nginx 容器内部，命令如下：

```
[root@node ~]# docker container cp index.html a6a22c7b6b1f:/usr/share/nginx/html
```

操作完成后，回到 master 节点，测试 Swarm 的负载均衡，使用命令查看 Nginx 首页，命令如下：

```
[root@master ~]# curl 192.168.200.17
web1.server
[root@master ~]# curl 192.168.200.17
web2.server
[root@master ~]# curl 192.168.200.17
web1.server
[root@master ~]# curl 192.168.200.17
web2.server
[root@master ~]# curl 192.168.200.17
web1.server
[root@master ~]# curl 192.168.200.17
web2.server
```

可以看到每次访问的 Nginx 页面会进行切换，验证负载均衡成功。

（8）Swarm 中的 Volume 服务

在 Swarm 集群中使用 Volume 服务，在 master 和 node 节点都创建一个 Volume，命令如下：

```
[root@master ~]# docker volume create --name testvolume
Testvolume
[root@node ~]# docker volume create --name testvolume
testvolume
```

查看创建的 Volume，命令如下：

```
[root@master ~]# docker volume ls
DRIVER              VOLUME NAME
```

```
local                   testvolume
[root@node ~]# docker volume ls
DRIVER                  VOLUME NAME
local                   testvolume
```

查看 Volume 的详细信息，命令如下：

```
[root@master ~]# docker volume inspect testvolume
[
    {
        "CreatedAt": "2021-09-28T04:24:07-04:00",
        "Driver": "local",
        "Labels": {},
        "Mountpoint": "/var/lib/docker/volumes/testvolume/_data",
        "Name": "testvolume",
        "Options": {},
        "Scope": "local"
    }
]
[root@node ~]# docker inspect testvolume
[
    {
        "CreatedAt": "2021-09-28T05:09:03-04:00",
        "Driver": "local",
        "Labels": null,
        "Mountpoint": "/var/lib/docker/volumes/testvolume/_data",
        "Name": "testvolume",
        "Options": null,
        "Scope": "local"
    }
]
```

在两个 Volume 路径下分别创建 index.html 文件，文件内容如下。

master 节点：

```
[root@master ~]# cd /var/lib/docker/volumes/testvolume/_data/
[root@master _data]# vi index.html
[root@master _data]# cat index.html
master_volume
```

node 节点：

```
[root@node ~]# cd /var/lib/docker/volumes/testvolume/_data/
[root@node _data]# vi index.html
[root@node _data]# cat index.html
node_volume
```

两个节点创建 index.html 文件后，创建新的服务并挂载 testvolume（以 Nginx 应用为例），命令如下：

```
[root@master ~]# docker service create --replicas 2 --mount type=volume,src=testvolume,dst=/usr/share/nginx/html --name test_nginx -p 80:80 nginx
tr4vuzejz2ip6fjfuuw99e4kg
overall progress: 2 out of 2 tasks
1/2: running   [==================================================>]
2/2: running   [==================================================>]
verify: Service converged
```

参数 src 可以写成 source；dst 表示容器内的路径，也可以写成 target。查看创建服务的详细信息，命令如下：

```
[root@master ~]# docker service ps test_nginx
ID              NAME           IMAGE          NODE      DESIRED STATE   CURRENT STATE
st612coh8u2i    test_nginx.1   nginx:latest   node      Running         Running 3 minutes ago
3e6dqgqb57hr    test_nginx.2   nginx:latest   master    Running         Running 3 minutes ago
```

查看挂载有没有成功，直接在 master 节点使用 curl 命令查看，如下所示：

```
[root@master ~]# curl 192.168.200.17
master_volume
[root@master ~]# curl 192.168.200.17
node_volume
[root@master ~]# curl 192.168.200.17
master_volume
[root@master ~]# curl 192.168.200.17
node_volume
[root@master ~]# curl 192.168.200.17
master_volum
[root@master ~]# curl 192.168.200.17
node_volume
```

可以看到两个节点的 Volume 卷挂载成功，且顺便验证一下 Swarm 的负载均衡功能。实验结束后，为了不影响后面的实验，将 test_nginx 删除，命令如下：

```
[root@master ~]# docker service rm test_nginx
test_nginx
```

4. Docker Swarm 集群实战

在上面的案例中，都是单独对一个 Nginx 服务进行集群部署，那如果要统一编排多个服务呢？Docker 中有个 Compose 服务，Compose 就是对单机进行统一编排的，它是通过 docker-compose.yaml 文件实现的，这里结合 Compose 和 Swarm 进行多服务的编排。

这里要部署的服务有 3 个（Nginx 服务、Visualizer 服务、Portainer 服务），都是集群 GUI 管理服务，Docker Service 部署的是单个服务，用户可以使用 docker stack 命令进行多服务编排部署。

因为在 Docker Swarm 集群搭建的时候，已经搭建了 Portainer 服务，所以先删除该服务，命令如下：

```
[root@master ~]# docker service rm q9staq87cu60 gdkr08yezhwq
q9staq87cu60
gdkr08yezhwq
```

接下来进行 3 个服务的编排部署，首先将提供的 dockersamples_visualizer_latest.tar 和 portainer_portainer_latest.tar 镜像上传至 master 节点的/root 目录下，并自行 load。

然后创建工作目录 testswarm，命令如下：

```
[root@master ~]# mkdir testswarm
[root@master ~]# cd testswarm/
```

在 testswarm 目录下创建 docker-compose.yaml，文件内容如下：

```
[root@master testswarm]# vi docker-compose.yaml
[root@master testswarm]# cat docker-compose.yaml
version: "3"
services:
  nginx:
    image: nginx:latest
    ports:
      - 8888:80
    deploy:
      mode: replicated
      replicas: 3

  visualizer:
    image: dockersamples/visualizer
    ports:
      - "8080:8080"
    volumes:
      - "/var/run/docker.sock:/var/run/docker.sock"
    deploy:
      replicas: 1
      placement:
        constraints: [node.role == manager]

  portainer:
    image: portainer/portainer
    ports:
```

```
      - "9000:9000"
    volumes:
      - "/var/run/docker.sock:/var/run/docker.sock"
    deploy:
      replicas: 1
      placement:
        constraints: [node.role == manager]
```

通过这个 yaml 文件部署服务，命令如下：

```
[root@master testswarm]# docker stack deploy -c docker-compose.yaml deploy_deamon
Creating network deploy_deamon_default
Creating service deploy_deamon_portainer
Creating service deploy_deamon_nginx
Creating service deploy_deamon_visualizer
```

从上面的执行过程可以看出这种创建方式会默认创建一个网络并使用它，创建的名字都是给定的前缀加上服务名，创建完成后，查看创建的服务，命令如下：

```
[root@master testswarm]# docker service ls
ID              NANE                      MODE         REPLICAS    IMAGE                              PORTS
wtk7a9hn47oh    deploy_deamon_nginx       replicated   3/3         nginx:latest                       *:8888->80/tcp
m45mf1wetfb9    deploy_deamon_portainer   replicated   1/1         portainer/portainer:latest         *:9000->9000/tcp
12wjjzaasxOz    deploy_deamon_visualizer  replicated   1/1         dockersamples/visualizer:latest    *:8080->8080/tcp
```

部署完成后，通过浏览器验证服务是否都部署成功，查看两个节点可以访问 http://IP:8888，如图 7-8 所示。

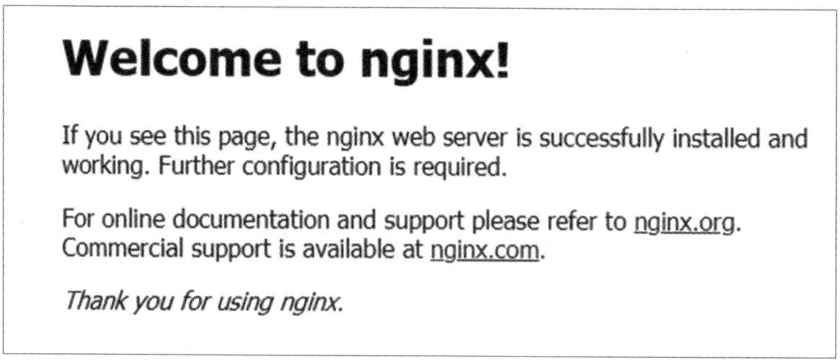

图 7-8 Nginx 首页

继续查看另外两个服务，可以使用 docker service ps 命令查看服务启动在哪个节点，通过浏览器访问 http://IP:8080，Visualizer 界面如图 7-9 所示。

通过浏览器访问 http://IP:9000，Portainer 服务界面如图 7-10 所示。

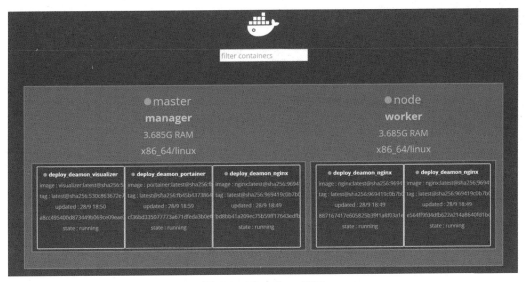

图 7-9　Visualizer 界面

图 7-10　Portainer 服务界面

至此，使用 Compose 和 Swarm 部署多服务案例验证成功。

归纳总结

通过本单元内容的学习，掌握了 Docker Swarm 集群服务的安装与基础使用。通过 Docker Swarm 服务实战案例的实践，更加深入地了解了 Docker Compose 和 Docker Swarm 的特性，Compose 只能编排部署在一台机器上，如果将 Compose 和 Swarm 结合，则可以将多服务编排部署到不同的节点上。

课后练习

一、判断题

1. Docker 容器技术的开发语言是 Go 语言。（   ）
2. Yaml 是一种非标记语言。（   ）

二、单项选择题

1. Docker Swarm 是 Docker 官方提供的一款（   ）工具。

   A. 虚拟化　　　　　B. 集群管理　　　　C. 监控管理　　　　D. 高可用

2. 下面（   ）是 Docker Swarm 中一个部署的最小单元。

   A. Stack　　　　　B. Yaml　　　　　　C. Task　　　　　　D. Service

三、多项选择题

Docker 架构包含了以下（   ）部分。

A. Containers　　　B. Images　　　　　C. Docker Daemon　D. REST API

技能训练

1. 安装 Docker 服务并搭建 Docker Swarm 集群。
2. 在 Docker Swarm 集群节点部署 Portainer 服务，并通过浏览器访问其页面。

单元 8　Kubernetes 集群服务

学习目标

通过本单元的学习，要求了解 Kubernetes 的概念、特性、集群架构和核心概念及部署方式和发展历程等。本单元培养读者快速掌握 Kubernetes 服务在工作中的实际运用，掌握 Kubernetes 集群的安装部署和应用体验、Kubernetes 服务运维与实战等技能；也培养读者提高自身综合能力，树立正确的职业素养。

8.1　认识 Kubernetes 集群服务

8.1.1　Kubernetes 介绍

容器集群管理系统 Kubernetes

1. Kubernetes 简介

Kubernetes 是一个全新的基于容器技术的分布式架构解决方案，是 Google 开源的一个容器集群管理系统。Kubernetes 这个名字源于希腊语，意为"舵手"或"飞行员"。K8S 这个缩写是因为 K 和 S 之间有 8 个字符的关系，所以 Kubernetes 简称 K8S。

Kubernetes 是一个一站式的完备的分布式系统开发和支撑平台，更是一个开放平台，对现有的编程语言、编程框架、中间件没有任何侵入性。

Kubernetes 提供了完善的管理工具，这些工具涵盖了开发、部署测试、运维监控在内的各个环节。

2. Kubernetes 的特性

（1）自我修复

在节点出现故障时，重新启动运行失败的容器，替换和重新部署，保证预期的副本数量；删除健康检查未通过的容器，并且在未准备好之前不会处理用户的请求，确保线上服务不中断。

（2）弹性伸缩

使用命令、UI 或者基于 CPU 使用情况自动快速扩容和缩容应用程序实例，保证应用业务高峰并发时的高可用性；业务低峰时回收资源，以最小成本运行服务。

（3）自动部署和回滚

K8S 采用滚动更新策略更新应用，一次更新一个 Pod，而不是同时删除所有 Pod，如果更新过程中出现问题，将回滚更改，确保升级不影响业务。

（4）服务发现和负载均衡

K8S 为多个容器提供一个统一的访问入口（内部 IP 地址和一个 DNS 名称），并且负载均衡关联的所有容器，使得用户无须考虑容器 IP 问题。

（5）机密和配置管理

管理机密数据和应用程序配置，而不需要把敏感数据暴露在镜像里，提高敏感数据的安全性，并可以将一些常用的配置存储在 K8S 中，方便应用程序的使用。

（6）存储编排

挂载外部存储系统，无论是来自本地存储、公有云，还是网络存储，都作为集群资源的一部分使用，极大地提高存储使用的灵活性。

（7）批处理

提供一次性任务、定时任务服务，满足批量数据处理和分析的场景。

3. Kubernetes 集群架构

Kubernetes 集群架构以及相关的核心组件如图 8-1 所示。

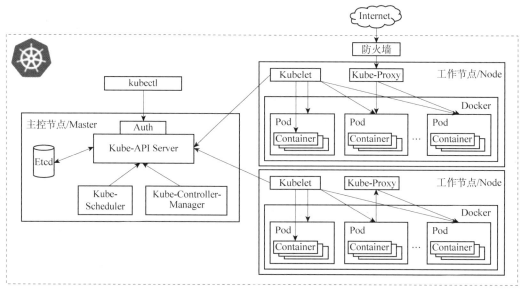

图 8-1 Kubernetes 集群架构以及相关的核心组件

一个 Kubernetes 集群一般包含一个 Master 节点和多个 Node 节点，一个节点可以看成是一台物理机或虚拟机。

Master 是 K8S 的集群控制节点，每个 K8S 集群里都需要有一个 Master 节点来负责整个集群的管理和控制，基本上 K8S 所有的控制命令都是发给它的，由它来负责具体的执行过程。Master 节点通常会占据一个独立的服务器，它太重要了，如果它不可用，那么所有的控制命令都将失效。

Master 节点上运行着以下关键组件。

（1）Kube-API Server

Kube-API Server 是集群的统一入口、各组件协调者，以 HTTP Rest 提供接口服务，所有对象资源的增、删、改、查和监听操作都交给 API Server 处理后再提交给 Etcd 存储。

（2）Kube-Controller-Manager

Kube-Controller-Manager 是 K8S 里所有资源对象的自动化控制中心，处理集群中的常规后台任务，一个资源对应一个控制器，而 Controller-Manager 就是负责管理这些控制器的。

（3）Kube-Scheduler

根据调度算法为新创建的 Pod 选择一个 Node 节点，可以部署在同一个节点上，也可以部署在不同的节点上。

（4）Etcd

Etcd 是一个分布式的、一致的 Key-Value 存储，主要用途是共享配置和服务发现、保存集群状态数据，比如 Pod、Service 等对象信息。

除了 Master 节点外，K8S 集群中的其他机器被称为 Node 节点，Node 节点是 K8S 集群中的工作负载节点，每个 Node 都会被 Master 分配一些工作负载，当某个 Node 宕机时，其上的工作负载会被 Master 自动转移到其他节点上去。

每个 Node 节点上都运行着以下关键组件：

① Kubelet。Kubelet 是 Master 在 Node 节点上的 Agent（代理），与 Master 密切协作，管理本机运行容器的生命周期，负责 Pod 对应的容器的创建、启停等任务，实现集群管理的基本功能。

② Kube-Proxy。在 Node 节点上实现 Pod 网络代理，实现 Kubernetes Service 的通信，维护网络规则和四层负载均衡工作。

③ Docker Engine。Docker 引擎，负责本机的容器创建和管理工作。Node 节点可以在运行期间动态增加到 K8S 集群中，前提是这个节点上已经正确安装、配置和启动了上述关键组件。在默认情况下 Kubelet 会向 Master 注册自己，一旦 Node 被纳入集群管理范围，Kubelet 就会定时向 Master 节点汇报自身的情况，例如，操作系统、Docker 版本、机器的 CPU 和内存情况，以及之前有哪些 Pod 在运行等，这样 Master 可以获知每个 Node 的资源使用情况，并实现高效均衡的资源调度策略。而某个 Node 超过指定时间不上报信息时，会被 Master 判定为"失联"，Node 的状态会被标记为不可用（Not Ready），随后 Master 会触发"工作负载大转移"的自动流程。

4. Kubernetes 的核心概念

（1）Pod

Pod 是 K8S 中最重要也是最基本的概念，Pod 是最小的部署单元，是一组容器的集合。每个 Pod 都由一个特殊的根容器（Pause 容器），以及一个或多个紧密相关的用户业务容器组成。

在 K8S 中，不会直接对容器进行操作，而是把容器包装成 Pod 再进行管理，Pod 在英文中的意思是豌豆荚，可以把容器想象成豌豆荚里的豆子，把一个或多个关系紧密的豆子包在一起就是豌豆荚（一个 Pod）。Pod 是运行服务的基础。

Pause 容器是 Pod 的根容器，以它的状态代表整个容器组的状态。K8S 为每个 Pod 都分配了唯一的 IP 地址，称为 Pod IP。Pod 里的多个业务容器共享 Pause 容器的 IP，共享 Pause 容器挂载的 Volume。

（2）Label

标签，附加到某个资源上，用于关联、查询和筛选对象。一个 Label 是一个 Key=Value 的键值对，Key 与 Value 由用户自己指定。Label 可以附加到各种资源上，一个资源对象可以定义任意数量的 Label，同一个 Label 也可以被添加到任意数量的资源上。

用户可以通过给指定的资源对象捆绑一个或多个不同的 Label 来实现多维度的资源分组管理功能，以便于灵活、方便地进行资源分配、调度、配置、部署等工作。

K8S 通过 Label Selector（标签选择器）来查询和筛选拥有某些 Label 的资源对象。Label Selector 有基于等式（name=label1）和基于集合［name in（label1，label2）］两种方式。

（3）ReplicaSet（RC）

ReplicaSet 用来确保预期的 Pod 副本数量，如果有过多的 Pod 副本在运行，系统就会停掉一些 Pod，否则系统就会再自动创建一些 Pod。

实际工作过程中很少单独使用 ReplicaSet，它主要被 Deployment 这个更高层的资源对象使用，从而形成一整套 Pod 创建、删除、更新的编排机制。

（4）Deployment

Deployment 用于部署无状态应用，Deployment 为 Pod 和 ReplicaSet 提供声明式更新，只需要在 Deployment 中描述想要的目标状态，Deployment 就会将 Pod 和 ReplicaSet 的实际状态改变到目标状态。

（5）Horizontal Pod Autoscaler（HPA）

HPA 为 Pod 横向自动扩容，也是 K8S 的一种资源对象。HPA 通过追踪分析 RC 的所有目标 Pod 的负载变化情况，来确定是否需要针对性地调整目标 Pod 的副本数量。

（6）Service

Service 定义了一个服务的访问入口，通过 Label Selector 与 Pod 副本集群之间"无缝对接"，定义了一组 Pod 的访问策略，防止 Pod 失联。

创建 Service 时，K8S 会自动为它分配一个全局唯一的虚拟 IP 地址，即 Cluster IP。服务发现就是通过 Service 的 Name 和 Service 的 Cluster IP 地址做一个 DNS 域名映射来解决的。

（7）Namespace

Namespace 即命名空间，Namespace 多用于实现多用户的资源隔离。Namespace 通过将集群内部的资源对象"分配"到不同的 Namespace 中，形成逻辑上分组的不同项目、小组或用户组。

K8S 集群在启动后，会创建一个名为 default 的 Namespace，如果不特别指明 Namespace，则创建的 Pod、RC、Service 都将被创建到 default 下。

当给每个用户创建一个 Namespace 来实现多用户的资源隔离时，还可以结合 K8S 的资源配额管理，限定不同用户能占用的资源，例如 CPU 使用量、内存使用量等。

8.1.2 Kubernetes 发展

1. 部署方式的发展

Kubernetes 的发展就是部署方式的发展，回顾这些年来的部署方式，从传统的硬件服务部署（传统部署时代）到虚拟机部署（虚拟化部署时代）再到容器化部署（容器部署时代），发展历程如图 8-2 所示。

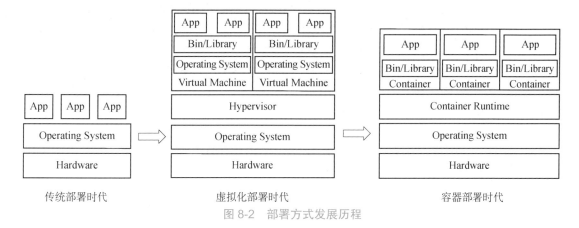

图 8-2　部署方式发展历程

（1）传统部署时代（Traditional Deployment）

早期，各个组织机构在物理服务器上运行应用程序，无法为物理服务器中的应用程序定义资源边界，这会导致资源分配问题。例如，如果在物理服务器上运行多个应用程序，则可能会出现一个应用程序占用大部分资源的情况，结果可能导致其他应用程序的性能下降。一种解决方案是在不同的物理服务器上运行每个应用程序，但是由于资源利用不足而无法扩展，并且维护许多物理服务器的成本很高。

（2）虚拟化部署时代（Virtualized Deployment）

作为解决方案，人们引入了虚拟化技术。虚拟化技术允许用户在单个物理服务器的 CPU 上运行多个虚拟机（VM）。虚拟化技术允许应用程序在 VM 之间隔离，并提供一定程度的安全性，因为一个应用程序的信息不能被另一应用程序随意访问。

虚拟化技术能够更好地利用物理服务器上的资源，并且因为可轻松地添加或更新应用程序而可以实现更好的可伸缩性、降低硬件成本等。

每个 VM 都是一台完整的计算机，在虚拟化硬件之上运行所有组件，包括其自己的操作系统。

（3）容器部署时代（Container Deployment）

容器类似于 VM，但是它们具有被放宽的隔离属性，可以在应用程序之间共享操作系统（OS）。因此，容器被认为是轻量级的。容器与 VM 类似，具有自己的文件系统、CPU、内存、进程空间等。由于它们与基础架构分离，因此可以跨云和 OS 发行版本进行移植。

容器因具有许多优势而变得流行起来。下面列出的是容器的一些优势。

① 敏捷应用程序的创建和部署：与使用 VM 镜像相比，提高了容器镜像创建的简便性和效率。

② 持续开发、集成和部署：通过快速简单的回滚（由于镜像不可变性），支持可靠且频繁的容器镜像构建和部署。

③ 关注开发与运维的分离：在构建/发布时而不是在部署时创建应用程序容器镜像，从而将应用程序与基础架构分离。

④ 可观察性：不仅可以显示操作系统级别的信息和指标，还可以显示应用程序的运行

状况和其他指标信号。

⑤ 跨开发、测试和生产的环境一致性：可以在便携式计算机上与在云中相同地运行。

⑥ 跨云和操作系统发行版本的可移植性：可在 Ubuntu、RHEL、CoreOS、本地、Google Kubernetes Engine 和其他任何地方运行。

⑦ 以应用程序为中心的管理：提高抽象级别，从在虚拟硬件上运行 OS 到使用逻辑资源在 OS 上运行应用程序。

⑧ 松散耦合、分布式、具有弹性的微服务：应用程序被分解成较小的独立部分，并且可以动态部署和管理，而不是在一台大型单机上整体运行。

⑨ 资源隔离：可预测的应用程序性能。

⑩ 资源利用：高效率和高密度。

2. Kubernetes 的发展

（1）2013 年——Docker 容器诞生

2013 年，以 Cloud Foundry 为代表的开源 PaaS 项目成了云计算领域的一股清流，PaaS 提供了一种"应用托管"的能力。

当时的虚拟机和云计算已经是比较普遍的技术了，主流用法就是租一批 AWS 或者 OpenStack 的虚拟机，然后用脚本或者手工的方式在机器上部署应用。

Cloud Foundry 这样的 PaaS 项目，核心组件就是一套打包和分发机制，会调用操作系统的 Cgroups 和 Namespace 机制为每个应用单独创建"沙盒"的隔离环境，然后在"沙盒"中运行这些进程，实现了多用户、批量、隔离运行的目的。这个"沙盒"，就是所谓的容器。

这一年还诞生了名为 dotCloud 的 Docker 公司，它也是追赶 PaaS 热潮中的一员。只不过，比起 Heroku、Pivotal、RedHat 等，dotCloud 公司显得太微不足道，主打产品跟主流的 Cloud Foundry 社区脱节，眼看就要"阵亡"的时候，dotCloud 公司决定开源自己的容器项目 Docker。

"容器"其实不是什么新鲜的东西，也不是 Docker 公司发明的，在当时最热的 PaaS 项目 Cloud Foundry 中，容器也只是底层的最不受关注的一部分。

然而短短几个月，Docker 就迅速崛起了，Cloud Foundry 等所有 PaaS 社区还没来得及成为对手，就已经被淘汰了。

Docker 项目和 Cloud Foundry 容器的大部分功能和实现原理是一样的，但是不一样的是"Docker 镜像"，它解决了环境打包的问题，它直接打包了应用运行所需要的整个操作系统，赋予了本地环境和云端环境调度一致的能力，避免了用户通过"试错"来匹配不同环境之间差异的痛苦过程，这也是 Docker 的精髓。

PaaS 变成了一套以 Docker 容器为技术核心，以 Docker 镜像为打包标准的"容器化"思路。2013 年年底，dotClound 公司正式改名为 Docker 公司。

（2）2014 年——Docker 发布编排工具

2014 年，Docker 发布 Docker Swarm 编排工具，以一个整体来对外提供集群管理功能，最大的亮点就是完全使用 Docker 项目原来的容器管理 API 来完成集群管理。

原来运行容器的命令如下：

```
docker run "容器"
```

Docker Swarm 工具只需要变成如下命令：

```
docker run -H "swarm集群API地址" "容器"
```

用户只需要使用原先的 Docker 指令创建一个容器，这个请求就会被 Swarm 拦截处理，通过具体的调度算法找到一个适合的 Docker Daemon。这种方式对已经熟悉 Docker 命令行的开发者们非常友好。

Swarm 可以使用 Docker 命令来管理集群下的 Docker 实例，但是 Swarm 并没有容器编排的能力。

接着，Docker 公司收购了 Fig 项目，后将其改名为 Compose。Fig 项目在开发者面前第一次提出了"容器编排"（Container Orchestration）概念。

在云计算领域，"编排"主要指用户如何通过某些工具或者配置来完成一组虚拟机以及关联资源的定义、配置、创建、删除等工作，然后由云计算平台按照这些指定的逻辑来完成的过程。

而在容器时代，"编排"就是对 Docker 容器的一系列定义、配置和创建动作的管理。

Docker 公司通过推出 Compose 来对容器进行编排。至此 Docker 公司达到了它的发展巅峰，几乎主宰了整个容器的生态。

（3）Docker 和 Mesosphere 公司的竞争

除了 Docker 生态外，Mesos 和背后的创建公司 Mesosphere 也是一个非常大的热点，Mesos 是大数据最受欢迎的资源管理项目，是跟 YARN 项目竞争的实力派对手。

大数据关注的计算密集型离线业务，其实不像 Web 服务那样适合用容器进行托管和扩容，也没有应用打包的强烈需要，所以 Hadoop、Spark 等项目现在也没在容器技术投入很大的精力，但是 Mesos 作为大数据套件之一，天生的两层调度机制让它非常容易从大数据领域独立出来去支持更广泛的 PaaS 业务，所以 Mesos 公司发布了 Marathon 项目，成了 Docker Swarm 的一个强有力的竞争对手。

虽然不能提供像 Swarm 那样的 Docker API，但是 Mesos 社区拥有一个非常大的竞争优势：超大规模集群管理经验，Mesos+Marathon 组合进化成了一个调度成熟的 PaaS 项目，同时能支持大数据业务。

（4）Docker 和 CoreOS 结束合作

CoreOS 是一个基础设施领域的创业公司，核心产品是一个定制化的操作系统，用户可以按照分布式集群的方式，管理所有安装了这个系统的节点，使用户在集群里部署应用像使用单机一样方便。

Docker 项目发布后，CoreOS 很快认识到可以把容器的概念无缝集成到自己的这套方案中，从而为用户提供更高层次的 PaaS 能力，所以 CoreOS 很早就成了 Docker 项目的贡献者，然而它们在 2014 年结束了合作，CoreOS 推出了自己的容器 Rocket（rkt），然而这个 rkt 容器完全被 Docker 公司压制了。

（5）OCI 标准的制定

由 CoreOS、Google、RedHat 等公司共同宣布，Docker 公司将 Libcontainer（容器运行时库）捐出，并改名为 RunC 项目，交由一个完全中立的基金会管理，并以 RunC 为依据共同制定一套容器和镜像的标准规范，这个标准规范叫 OCI（Open Container Initiative）。

制定这个标准的本意在于将容器运行和镜像的实现从 Docker 项目中完全剥离出来,以此来压制 Docker 公司一家独大的现状,同时也为不依赖 Docker 项目构建平台层能力提供了可能。不过这个标准并没有改变 Docker 在容器领域一家独大的现状。

(6) 2014 年——Kubernetes 诞生

2014 年 6 月,基础设施领域的领先者 Google 正式宣告了 Kubernetes 项目的诞生(Borg 的开源版本),如同 Docker 横空出世一样,它再一次改变了容器市场的格局。微软、RedHat、IBM、Docker 均加入 Kubernetes 社区。

(7) 2015 年——CNCF 基金会成立

Google、RedHat 等公司为了对抗以 Docker 公司为核心的容器商业生态,一起成立了 CNCF(Cloud Native Computing Foundation)基金会,希望以 Kubernetes 为基础,建立一个由开源基础设施领域厂商主导、按照独立基金会方式运营的平台社区,来对抗以 Docker 公司为核心的容器商业生态,简单地说就是打造一个围绕 Kubernetes 项目的"护城河"。

成立 CNCF 的目的是以开源的 K8S 为基础,使得 K8S 能够在容器编排方面覆盖更多的场景,提供更强的能力。K8S 必须面临 Swarm 和 Mesos 的挑战。Swarm 的强项是和 Docker 生态的天然无缝集成,而 Mesos 的强项是大规模集群的管理和调度。Kubernetes 的团队规模很小,投入的工程能力有限,RedHat 在这时候和 Google 联盟,正式开启了容器编排"三国鼎立"的局面。

Kubernetes 是 Google 容器化基础设施领域多年来实践经验的沉淀和升华,在 Github 上的各项指标一路飙升,将 Swarm 项目远远地甩在了后边。

同年,Kubernetes 发布 Helm 软件包管理系统、Kubeadm 安装工具,并发布 Mikibube 等一系列更新操作。

CNCF 社区迅速增加了 Prometheus、Fluentd、OpenTracing、CNI 等一系列容器生态的知名工具和项目,大量的公司和创建团队将注意力投向 CNCF 社区而不再是 Docker 公司。

(8) 2016 年——Docker 放弃 Swarm

2016 年,面对 CNCF 基金会的竞争优势,Docker 公司放弃现有的 Swarm 项目,将容器编排和集群管理功能内置到 Docker 中。然而这种改变带来的技术复杂度和维护难度,给 Docker 项目造成了非常不利的局面。

同年,Kuberntes 支持 OpenApi,给开发人员定制化提供更大的灵活性。不同于 Docker 公司,Kubernetes 推进"民主化"架构,从 API 到容器运行的每一层,都给开发者暴露出了可扩展的插件机制,鼓励用户通过代码的方式介入每一个阶段。

Kubernetes 项目的这次变革非常有效,很快在整个容器社区中催生出了大量基于 Kubernetes API 和扩展接口的二次创新产品,具体如下:

① 热度极高的 Istio 微服务治理工具。

② 应用部署框架 Operator。

③ Rook 开源创业项目,把 Ceph 重量级产品封装成了简单易用的容器存储插件。

Docker 公司在 Kubernetes 社区崛起和壮大后,败下阵来。

(9) 2017 年——Docker 将 Containerd 捐献给 CNCF 社区

2017 年,Docker 公司将 Containerd 捐献给 CNCF 社区,Docker 公司宣布将 Docker 项目改名为 Moby,交给社区自行维护,而 Docker 公司的商业产品还占有 Docker 这个注册商标。

同年 10 月，Docker 宣布将在自己的主打产品 Docker 企业版中内置 Kubernetes 项目，持续了两年的容器编排之争终于落下帷幕。

8.2 使用 Kubernetes 集群服务

原生 Kubernetes
云平台部署（1）

1. 环境准备

（1）规划节点

Kubernetes 实验节点规划如表 8-1 所示。

表 8-1 Kubernetes 实验节点规划

IP 地址	主机名	节点
192.168.200.19	master	Kubernetes 实验节点

（2）基础准备

准备好 VMware Workstation 工具和 CentOS7.5 版本虚拟机，使用 CentOS7.5 虚拟机快照，克隆一台虚拟机，云主机配置自定义，推荐配置 4vCPU、12GB 内存、100GB 硬盘，网络使用 NAT 模式，master 节点 IP 设置为 192.168.200.19，推荐设置主机密码为 000000。请自行将虚拟机配置好 IP、安装 Docker 服务，关闭 SELinux 和防火墙服务。

2. Kubernetes 部署

（1）部署架构

K8S 集群中有管理节点与工作节点两种类型，部署架构如图 8-3 所示。

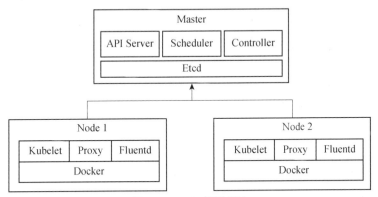

图 8-3 K8S 部署架构

从图 8-3 可以看到，一般 K8S 集群会部署一个 Master 节点和若干个 Node 节点，Master 节点主要负责 K8S 集群管理，集群中各节点间的信息交互、任务调度，还负责容器、Pod、Namespaces、PV 等生命周期的管理。Node 节点主要为容器和 Pod 提供计算资源，Pod 及容器全部运行在 Node 节点上，Node 节点通过 Kubelet 服务与管理节点通信以管理容器的生命周期，并与集群中其他节点进行通信。

（2）环境准备

使用 VMware Workstation 进行实验，考虑到 PC 的配置，使用单节点安装 Kubernetes

服务，将 Master 节点和 Node 节点安装在一个节点上，配置如表 8-2 所示。

表 8-2 环境准备

节点角色	主机名	内存	硬盘	IP 地址
Master/Node	master	12GB	100GB	192.168.200.19

此次安装 Kubernetes 服务的系统为 CentOS7.5-1804；Docker 版本为 docker-ce-19.03.13；Kubernetes 版本为 1.18.1。

（3）安装 Docker 并部署 Harbor

① 基础设置。使用远程连接工具 SSH 到 192.168.200.19 节点，修改主机名为 master，并关闭该节点的 SELinux 和防火墙服务，关于如何修改主机名和关闭 SELinux 及防火墙服务，此处不再赘述，修改完之后，如下所示：

```
[root@master ~]# hostnamectl
   Static hostname: master
         Icon name: computer-vm
           Chassis: vm
        Machine ID: 32bedca2aca04e2b82c5e40e1040cf3f
           Boot ID: 835ebc44dfc0421b9a66ac20590402b4
    Virtualization: vmware
  Operating System: CentOS Linux 7 (Core)
       CPE OS Name: cpe:/o:centos:centos:7
            Kernel: Linux 3.10.0-862.el7.x86_64
      Architecture: x86-64
[root@master ~]# getenforce
Disabled
[root@master ~]# systemctl status firewalld
• firewalld.service - firewalld - dynamic firewall daemon
   Loaded: loaded (/usr/lib/systemd/system/firewalld.service; disabled; vendor preset: enabled)
   Active: inactive (dead)
     Docs: man:firewalld(1)
```

可以看到 SELinux 处于 Disabled 状态，防火墙也处于 inactive。

② 删除防火墙规则。将 master 节点的 iptables 规则清除，命令如下：

```
[root@master ~]# iptables -F
[root@master ~]# iptables -X
[root@master ~]# iptables -Z
[root@master ~]# /usr/sbin/iptables-save
```

③ 配置 YUM 源。将提供的 CentOS-7-x86_64-DVD-1804.iso 和 chinaskills_cloud_paas.iso 镜像文件上传至 master 节点的/root 目录下，把 CentOS-7-x86_64-DVD-1804.iso 挂载到/opt/centos 目录下（/opt/centos 目录需自行创建），然后将 chinaskills_cloud_paas.iso 挂载到/mnt 下，并将/mnt 目录下的所有文件及软件包复制至/opt 目录下，做完上述操作后，创建本地

YUM 源文件 local.repo，文件内容如下：

```
[root@master ~]# cat /etc/yum.repos.d/local.repo
[k8s]
name=k8s
baseurl=file:///opt/kubernetes-repo
gpgcheck=0
enabled=1
[centos]
name=centos
baseurl=file:///opt/centos
gpgcheck=0
enabled=1
```

④ 安装 Docker 及 Harbor。进入/opt 目录，安装 Docker 服务和 Harbor 镜像仓库的操作均已整理成脚本 k8s_harbor_install.sh，在 master 节点执行脚本 k8s_harbor_install.sh 即可完成 Docker 与 Harbor 仓库的搭建，命令如下：

原生 Kubernetes 云平台部署（2）

```
[root@master opt]# ./k8s_harbor_install.sh
```

等待一段时间后，Docker 服务与 Harbor 仓库安装完毕，登录 http://192.168.200.19 访问 Harbor 仓库界面（登录用户为 admin；密码为 Harbor12345），如图 8-4 所示。

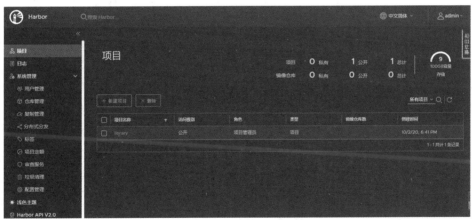

图 8-4　Harbor 仓库界面

至此，Docker 环境与 Harbor 镜像仓库安装完毕，接下来需要将部署 Kubernetes 用到的镜像上传至 Harbor 镜像仓库中。

（4）上传镜像

上传镜像的操作前先生成脚本 k8s_image_push.sh，执行该脚本，再上传镜像，命令如下：

```
[root@master opt]# ./k8s_image_push.sh
输入镜像仓库地址(不加http/https): 192.168.200.19
输入镜像仓库用户名: admin
```

```
输入镜像仓库用户密码：Harbor12345
您设置的仓库地址为：192.168.200.19,用户名：admin,密码：xxx
是否确认(Y/N)：Y
```

进入 Harbor 仓库 library 项目查看镜像列表，如图 8-5 所示。

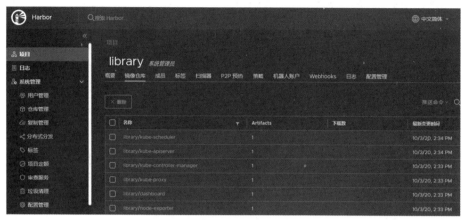

图 8-5　Harbor 仓库列表

镜像已上传至 Harbor 仓库，接下来部署 Kubernetes。

（5）部署 Kubernetes

部署 Kubernetes 服务的操作步骤也已经生成了脚本，分别为 k8s_master_install.sh 和 k8s_node_install.sh。因为此处将 Master 节点和 Node 节点合并了，只需要执行 k8s_master_install.sh 一个脚本，命令如下：

```
[root@master opt]# ./k8s_master_install.sh
```

等待一段时间后，Kubernetes 安装完毕。在平时的工作中，安装 Kubernetes 时出于安全考虑，默认配置下 Kubernetes 不会将 Pod 调度到 Master 节点上。但此时因为资源的管理，将 Master 和 Node 两个节点都安装在了一起，所以需要在 Master 节点执行删除污点操作，这样 Master 节点也能当作 Node 节点使用了。删除污点操作的命令如下：

```
# kubectl taint node master node-role.kubernetes.io/master-
```

该命令已写在安装脚本中。

在 Master 节点查看节点的状态，命令如下：

```
[root@master opt]# kubectl get nodes
NAME     STATUS   ROLES    AGE   VERSION
master   Ready    master   98d   v1.18.1
```

至此，Kubernetes 服务安装完毕，接下来了解 Kubernetes 有哪些常用的命令、这些命令分别用来做什么。

（6）Kubernetes 命令介绍

① kubectl 命令语法如下：

```
# kubectl [command] [TYPE] [NAME] [flags]
```

命令语法解析：
- command：子命令。
- TYPE：资源类型。
- NAME：资源名称。
- flags：命令参数。

② kubectl 命令选项如下：
- --alsologtostderr[=false]：同时输出日志到标准错误控制台和文件。
- --api-version=""：和服务端交互使用的 API 版本。
- --certificate-authority=""：用以进行认证授权的.cert 文件路径。
- --client-certificate=""：TLS 使用的客户端证书路径。
- --client-key=""：TLS 使用的客户端密钥路径。
- --cluster=""：指定使用的 kubeconfig 配置文件中的集群名。
- --context=""：指定使用的 kubeconfig 配置文件中的环境名。
- --insecure-skip-tls-verify[=false]：如果为 true，将不会检查服务器凭证的有效性，这会导致 HTTPS 链接变得不安全。
- --kubeconfig=""：命令行请求使用的配置文件路径。
- --log-backtrace-at=：0：当日志长度超过定义的行数时，忽略堆栈信息。
- --log-dir=""：如果不为空，将日志文件写入此目录。
- --log-flush-frequency=5s：刷新日志的最大时间间隔。
- --logtostderr[=true]：输出日志到标准错误控制台，不输出到文件。
- --match-server-version[=false]：要求服务端和客户端版本匹配。
- --namespace=""：如果不为空，命令将使用此 Namespace。
- --password=""：API Server 进行简单认证使用的密码。
- -s，--server=""：Kubernetes API Server 的地址和端口号。
- --stderrthreshold=2：高于此级别的日志将被输出到错误控制台。
- --token=""：认证到 API Server 使用的令牌。
- --user=""：指定使用的 kubeconfig 配置文件中的用户名。
- --username=""：API Server 进行简单认证使用的用户名。
- --v=0：指定输出日志的级别。
- --vmodule=：指定输出日志的模块。

③ 常用命令如下：
- kubectl annotate：更新资源的注解。
- kubectl api-versions：以"组/版本"的格式输出服务端支持的 API 版本。
- kubectl apply：通过文件名或控制台输入，对资源进行配置。
- kubectl attach：连接到一个正在运行的容器。
- kubectl autoscale：对 Replication Controller 进行自动伸缩。
- kubectl cluster-info：输出集群信息。
- kubectl config：修改 kubeconfig 配置文件。
- kubectl create：通过文件名或控制台输入创建资源。

- kubectl delete：通过文件名、控制台输入、资源名或者 Label Selector 删除资源。
- kubectl describe：输出指定的一个或多个资源的详细信息。
- kubectl edit：编辑服务端的资源。
- kubectl exec：在容器内部执行命令。
- kubectl expose：输入 rc、svc 或 Pod，并将其暴露为新的 Kubernetes Service。
- kubectl get：输出一个或多个资源。
- kubectl label：更新资源的 Label。
- kubectl logs：输出 Pod 中一个容器的日志。
- kubectl namespace：（已停用）设置或查看当前使用的 Namespace。
- kubectl patch：通过控制台输入更新资源中的字段。
- kubectl port-forward：将本地端口转发到 Pod。
- kubectl proxy：为 Kubernetes API Server 启动代理服务器。
- kubectl replace：通过文件名或控制台输入替换资源。
- kubectl rolling-update：对指定的 Replication Controller 执行滚动升级。
- kubectl run：在集群中使用指定的镜像启动容器。
- kubectl scale：为 Replication Controller 设置新的副本数。
- kubectl version：输出服务端和客户端的版本信息。

④ 常用命令体验。获取 Kubernetes 集群下 Namespace 的信息，命令如下：

```
[root@master ~]# kubectl get namespace
NAME                STATUS   AGE
default             Active   102d
kube-node-lease     Active   102d
kube-public         Active   102d
kube-system         Active   102d
```

kubectl get namespace 命令还可以写成 kubectl get ns 命令，效果是一样的。

查看命名空间下的所有信息，命令如下：

```
[root@master ~]# kubectl get all
NAME                 TYPE        CLUSTER-IP   EXTERNAL-IP   PORT(S)   AGE
service/kubernetes   ClusterIP   10.96.0.1    <none>        443/TCP   102d
```

kubectl get all 这条命令等同于"kubectl get pod, svc, deployment, rs"。all 包括所有 pod、svc、deployment 和 rs。因为在默认的 Namespace 下没有其他的 pod、deployment 和 rs，所以就只显示了 Service。

可以查看某个 Namespace 中的所有信息，比如查看 kube-system 下所有的信息，命令如下：

```
[root@master ~]# kubectl get all -n kube-system
NAME                              READY   STATUS    RESTARTS   AGE
pod/coredns-666749ccb9-bw9m4      1/1     Running              102d
pod/coredns-666749ccb9-rvxjg      1/1     Running              102d
```

```
pod/etcd-master                           1/1      Running    102d
pod/ kube-apiserver-master                1/1      Running    102d
pod/ kube-controller-manager-master       1/1      Running    102d
pod/ kube-flannel-ds-zzsdx                1/1      Running    102d
pod/ kube-proxy-8dbbb                     1/1      Running    102d
pod/ kube-scheduler-master                1/1      Running    102d
NAME                 TYPE        CLUSTER-IP   EXTERNAL-IP  PORT(S)              AGE
service/kube-dns  clusterIP  10.96.0.10    <none>      53/UDP,53/TCP ,9153/TCP
102d
  NANE           DESIRED  CURRENT  READY  UP-TO-DATE  AVAILABLE  NODE SELECTOR      AGE
daemonset.apps/ kube-flannel-ds    1    1    1    1    1    <none>         102d
daemonset.apps/ kube-proxy         1    1    1    1    1    kubernetes.io/os=
linux 102d
  NANE                        READY  UP-TO-DATE  AVAILABLE   AGE
deployment.apps/coredns       2/2       2           2        102d
  NANE                              DESIRED  CURRENT  READY AGE
replicaset.apps/coredns-666749ccb9   2       2        2     2   102d
```

也可以查看某项资源,比如查看 Pod,命令如下:

```
[root@master ~]# kubectl get pods
No resources found in default namespace.
```

可以发现当前的 default 这个 Namespace 下没有 Pod,可以指定其他的 Namespace 来查看该命名空间下的 Pod,命令如下:

```
[root@master ~]# kubectl get pods -n kube-system
NAME                                  READY    STATUS     RESTARTS   AGE
coredns-666749ccb9-bw9m4              1/1      Running    2          102d
coredns-666749ccb9-rvxjg              1/1      Running    2          102d
etcd-master                           1/1      Running    2          102d
kube-apiserver-master                 1/1      Running    2          102d
kube-controller-manager-master        1/1      Running    3          102d
kube-flannel-ds-zzsdx                 1/1      Running    3          102d
kube-proxy-8dbbb                      1/1      Running    2          102d
kube-scheduler-master                 1/1      Running    3          102d
```

还可以在命令后面加上 -o wide 来查看更多的信息,读者可以自行输入命令尝试。关于创建资源的命令如下:

```
# 通过文件创建一个 Deployment
# kubectl create -f /path/to/deployment.yaml
# 不过现在一般常用下面的命令来创建资源
# kubectl apply -f /path/to/deployment.yaml
# 通过 kubectl 命令直接创建
# kubectl run nginx_app --image=nginx:1.9.1 --replicas=3
```

关于 Kubernetes 中常用的命令就介绍到这里，接下来通过几个小案例更深入地了解 Kubernetes 的命令、特性、作用与功能。

2. Kubernetes 体验

一旦运行了 Kubernetes 集群，就可以在其上部署容器化的应用程序了。为此，需要创建 Kubernetes Deployment。Deployment 指挥 Kubernetes 如何创建和更新应用程序的实例。创建 Deployment 后，Kubernetes Master 将应用程序实例调度到集群中的各个节点上。

在 Kubernetes 上部署第一个应用前，还需要掌握几个概念。在 Kubernetes 中不会直接对容器进行操作，在 Kubernetes 中，Pod 是最小的可部署单元。除了要了解 Pod 外，还需要掌握 ReplicaSet、Deployment、Service 的概念。K8S 常用概念关系如图 8-6 所示。

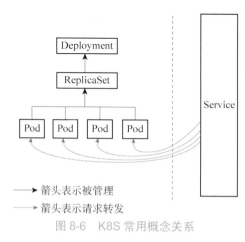

图 8-6　K8S 常用概念关系

① Pod。Pod 是一个或多个容器的组合，这些容器共享存储、网络和命名空间，以及运行的规范。

② ReplicaSet。在介绍 ReplicaSet 之前，先要介绍一下 Replication Controller。Replication Controller 的作用是确保 Pod 以指定的副本个数运行。

ReplicaSet 是 Replication Controller 的升级版。ReplicaSet 和 Replication Controller 之间的唯一区别是对 Selector（选择器）的支持。Replication Controller 只支持基于等式的 Selector，但 ReplicaSet 还支持新的、基于集合的 Selector。

在 Yaml 文件中通过 spec.replicas 声明 Pod 的副本数。

③ Deployment。Deployment 用于管理 Pod、ReplicaSet，可实现滚动升级和回滚应用、扩容和缩容。

④ Service。试想一个问题，ReplicaSet 定义了 2 个 Pod，当其中一个 Pod 由于某种原因停止时，ReplicaSet 会新建一个 Pod，以确保运行中的 Pod 数量始终是 2。但每个 Pod 都有自己的 IP，前端请求不知道这个新 Pod 的 IP，那前端的请求如何被发送到新 Pod 中呢？

答案是使用 Service。K8S 的 Service 定义了一个服务的访问入口地址，前端的应用通过这个入口地址访问其背后的一组由 Pod 副本组成的集群实例，来自外部的访问请求被负载均衡到后端的各个容器应用上。Service 与其后端 Pod 副本集群之间则是通过 Label Selector 实现关联的，通俗地讲就是前端请求不是直接发送给 Pod 的，而是发送到 Service，

Service 再将请求转发给 Pod。

总结：Pod 被 ReplicaSet 管理，ReplicaSet 控制 Pod 的数量；ReplicaSet 被 Deployment 管理，Deployment 控制 Pod 应用的升级、回滚，当然也能控制 Pod 的数量。Service 提供一个统一固定入口，负责将前端请求转发给 Pod。

（1）Kubernetes 部署应用初体验

使用 K8S 命令部署一个 Nginx 应用，命令如下：

原生 Kubernetes 容器云平台基本测试使用

```
[root@master ~]# kubectl run nginx --image=192.168.200.19/library/nginx: latest
pod/nginx created
```

在之前的版本中，使用 kubectl run 命令时，系统会先创建一个 Deployment，然后再由控制器去创建 Nginx 的 Pod，但是在新版本中，使用 kubectl run 命令只会单独创建一个 Pod，并不会创建 Deployment。可以使用命令查看 Nginx 的 Pod 和 Deployment，命令如下：

```
[root@master ~]# kubectl get pods
NAME    READY   STATUS    RESTARTS   AGE
nginx   1/1     Running   0          14s
[root@master ~]# kubectl get deployment
No resources found in default namespace.
```

除了 Pod 被创建之外，Deployment 并没有被创建。如果要访问 Nginx 服务，则只能在宿主机进行访问，外部访问不到 Nginx 服务。

查看 Pod 的 IP 地址，命令如下：

```
[root@master ~]# kubectl get pods -o wide
NAME    READY   STATUS    RESTARTS   AGE   IP             NODE     NOMINATED NODE   READINESS GATES
nginx   1/1     Running   0          12m   10.244.0.34    master   <none>           <none>
```

通过 kubectl get pods -o wide 命令获取的 Pod 的 IP 地址是一个虚拟的 IP 地址，外部是访问不到的。在 Master 节点使用 curl 命令查看 Pod 地址，命令如下：

```
[root@master ~]# curl http://10.244.0.34
<!DOCTYPE html>
<html>
<head>
<title>Welcome to nginx!</title>
<style>
    body {
        width: 35em;
        margin: 0 auto;
        font-family: Tahoma, Verdana, Arial, sans-serif;
    }
</style>
</head>
```

```
<body>
<h1>Welcome to nginx!</h1>
<p>If you see this page, the nginx web server is successfully installed and
working. Further configuration is required.</p>

<p>For online documentation and support please refer to
<a href="http://nginx.org/">nginx.org</a>.<br/>
Commercial support is available at
<a href="http://nginx.com/">nginx.com</a>.</p>

<p><em>Thank you for using nginx.</em></p>
</body>
</html>
```

部署的 Nginx 应用只能在内部访问，那么怎么样才能实现外部访问呢？在上面介绍了，Service 是 Kubernetes 中用于服务发现的，在下面的案例中，将会介绍怎么使用 Deployment 和 Service 部署应用与服务发现。

（2）使用 Deployment 部署应用

使用 Deployment 的方式部署 Nginx 应用，可以使用如下命令：

```
[root@master ~]# kubectl create deployment nginx --image=192.168.200.19/library/nginx:latest
deployment.apps/nginx created
```

可以看到一个名叫 nginx 的 Deployment 被创建，查看 Deployment 列表，命令如下：

```
[root@master ~]# kubectl get deployment
NAME    READY   UP-TO-DATE   AVAILABLE   AGE
nginx   1/1     1            1           35s
```

继续查看 Pod，命令如下：

```
[root@master ~]# kubectl get pods
NAME                       READY   STATUS    RESTARTS   AGE
nginx                      1/1     Running   0          42m
nginx-86d8d488f6-pkmcb     1/1     Running   0          81s
```

可以看到创建 Deployment 时会自动创建一个 Pod，这个时候，Nginx 应用也只能内部访问，不能通过外部访问，还需要进行 Service 服务发现或者叫端口开放，才能进行外部访问，Service 服务发现的命令如下：

```
[root@master ~]# kubectl expose deployment nginx --port=80 --type=NodePort
service/nginx exposed
```

做完该操作后，查看 Service，命令如下：

```
[root@master ~]# kubectl get svc
NAME         TYPE        CLUSTER-IP   EXTERNAL-IP   PORT(S)   AGE
kubernetes   ClusterIP   10.96.0.1    <none>        443/TCP   99d
```

```
nginx          NodePort     10.99.28.34      <none>              80:30209/TCP       3s
```

可以看到 Nginx 的 80 端口被映射到 30209 端口，这时就可以从外部访问 Nginx 应用了。使用 PC 的浏览器访问 http://192.168.200.19:30209，如图 8-7 所示。

图 8-7 Nginx 首页

在这个案例中，使用 Deployment 管理了 Pod，并使用了 Service 用户服务发现，没有使用 ReplicaSet 控制器定义副本的数量。在实际的工作中，不会使用执行命令的方式部署应用，一般在 Kubernetes 中，都会编写 Yaml 模板文件进行编排部署应用，下面通过一个简单的例子，尝试使用 Yaml 文件编排部署应用。

（3）使用 Yaml 模板部署应用

使用 Yaml 文件来部署 Nginx 服务，要求使用 Deployment 管理 Pod，使用 ReplicaSet 控制器定义副本的数量为 2。首先创建 Yaml 文件，命令如下：

```
[root@master ~]# vi nginx-deployment.yaml
```

nginx-deployment.yaml 文件的内容如下所示：

```
# API 版本号
apiVersion: apps/v1
# 类型,如:Pod/ReplicationController/Deployment/Service/Ingress
kind: Deployment
metadata:
  # Kind 的名称
  name: nginx-app
spec:
  selector:
    matchLabels:
      # 容器标签的名字,发布 Service 时,selector 需要和这里对应
      app: nginx
  # 部署的实例数量
  replicas: 2
template:
  metadata:
    labels:
      app: nginx
  spec:
```

```
        # 配置容器,数组类型,说明可以配置多个容器
        containers:
        # 容器名称
        - name: nginx
          # 容器镜像
          image: 192.168.200.19/library/nginx:latest
          # 只有镜像不存在时,才会进行镜像拉取
          imagePullPolicy: IfNotPresent
          ports:
          # Pod 端口
          - containerPort: 80
```

编辑好 Yaml 文件后,运行该文件,命令如下:

```
[root@master ~]# kubectl apply -f nginx-deployment.yaml
deployment.apps/nginx-app created
```

查看运行的 Pod,命令如下(此处已将上面案例中创建的 Pod 和 Deployment 删除):

```
[root@master ~]# kubectl get pods
NAME                          READY   STATUS    RESTARTS   AGE
nginx-app-7bd9c4cb95-dkw8r    1/1     Running   0          48s
nginx-app-7bd9c4cb95-dl5b7    1/1     Running   0          48s
```

查看运行的 Deployment,命令如下:

```
[root@master ~]# kubectl get deployment
NAME        READY   UP-TO-DATE   AVAILABLE   AGE
nginx-app   2/2     2            2           2m25s
```

可以看到当前 Pod 中显示 2 个正在运行的 Pod,Deployment 中的 Nginx 数量也是 2/2。但是此时只使用了 Deployment 管理 Pod,然后使用 ReplicaSet 控制器设置副本的数量为 2,此时还不能通过外部网络访问 Nginx 应用,还需要做 Service 发现,创建 Service 的 Yaml 文件,命令如下:

```
[root@master ~]# vi nginx-service.yaml
```

nginx-service.yaml 文件的内容如下所示:

```
[root@master ~]# cat nginx-service.yaml
apiVersion: v1
# 类型,如: Pod/ReplicationController/Deployment/Service/Ingress
kind: Service
metadata:
  # Kind 的名称
  name: nginx-service
spec:
  selector:
```

```
    app: nginx
  ports:
  - port: 80
    protocol: TCP
    targetPort: 80
  type: NodePort
```

执行 nginx-service.yaml 文件，命令如下：

```
[root@master ~]# kubectl apply -f nginx-service.yaml
service/nginx-service created
```

查看创建的 Service，命令如下：

```
[root@master ~]# kubectl get svc
NAME             TYPE        CLUSTER-IP        EXTERNAL-IP   PORT(S)        AGE
kubernetes       ClusterIP   10.96.0.1         <none>        443/TCP        101d
nginx-service    NodePort    10.102.105.232    <none>        80:31526/TCP   19s
```

可以通过宿主机 IP 地址:31526 访问 Nginx，如图 8-8 所示。

图 8-8　Nginx 首页（1）

模拟其中一个 Nginx 服务挂掉，看是否会影响服务。首先查看 Pod，命令如下：

```
[root@master ~]# kubectl get pods
NAME                           READY   STATUS    RESTARTS   AGE
nginx-app-7bd9c4cb95-dkw8r     1/1     Running   0          15m
nginx-app-7bd9c4cb95-dl5b7     1/1     Running   0          15m
```

当前存在 2 个 Pod，删除其中一个 Pod，命令如下：

```
[root@master ~]# kubectl delete pods nginx-app-7bd9c4cb95-dkw8r
pod "nginx-app-7bd9c4cb95-dkw8r" deleted
```

删除完之后，使用网页查看 Nginx 服务的首页，服务不受影响，继续查看 Pod 数量，命令如下：

```
[root@master ~]# kubectl get pods
NAME                           READY   STATUS    RESTARTS   AGE
nginx-app-7bd9c4cb95-dl5b7     1/1     Running   0          15m
nginx-app-7bd9c4cb95-zsw4n     1/1     Running   0          4s
```

可以看到 Pod 的数量是 2，因为在启动应用时，使用 ReplicaSet 控制器定义了副本数量为 2，所以如果有一个 Nginx 服务挂掉，会自动切换 Pod，而且后台也会重新生成新的 Pod，保证副本数一直是 2。

通过这个实验案例，相信读者对 Kubernetes 的使用有了一定的认识，也掌握了 Yaml 文件的基本编写方法，对 Pod、Deployment、ReplicaSet、Service 也有了更深的认识。接下来通过一个 Kubernetes 的实战案例，让读者更进一步地掌握 Kubernetes 服务的使用。

3. Kubernetes 实战

本实战案例拟使用 Kubernetes 平台部署一个 WordPress 博客系统，具体配置与介绍如下：
① 需要用到两个容器镜像，分别为 WordPress:4.8-apache 和 MySQL:5.6。
② 使用 NFS 创建 PV 作为数据库与 WordPress 的后端存储。
③ 使用 Secret 生成器管理密码。
④ 最后使用 Yaml 文件编排部署博客系统。

（1）安装与配置 NFS
在 master 节点上安装 NFS 所需要的软件及依赖，命令如下：

原生 Kubernetes 容器云平台应用部署（1）

```
[root@master ~]# yum install -y nfs-utils rpcbind
```

安装完之后，启动 RPC 与 NFS 服务，命令如下：

```
[root@master ~]# systemctl start rpcbind
[root@master ~]# systemctl enable rpcbind
[root@master ~]# systemctl start nfs
[root@master ~]# systemctl enable nfs
```

创建两个目录，作为 NFS 的共享目录，然后赋予权限，命令如下：

```
[root@master ~]# mkdir -p /home/pvdata/mysql
[root@master ~]# mkdir -p /home/pvdata/wordpress
[root@master ~]# chmod 777 /home/pvdata/mysql
[root@master ~]# chmod 777 /home/pvdata/wordpress
```

修改 NFS 配置文件/etc/exports，将创建的两个目录进行共享，命令如下：

```
[root@master ~]# vi /etc/exports
[root@master ~]# cat /etc/exports
#在/etc/exports 配置文件中,添加如下两行
/home/pvdata/mysql 192.168.200.0/24(rw,no_root_squash,no_all_squash,sync,anonuid=501,anongid=501)
/home/pvdata/wordpress 192.168.200.0/24(rw,no_root_squash,no_all_squash,sync,anonuid=501,anongid=501)
```

生效 NFS 配置文件，并查看分享的目录，命令如下：

```
[root@master ~]# exportfs -r
[root@master ~]# showmount -e 192.168.200.19
Export list for 192.168.200.19:
```

```
/home/pvdata/wordpress 192.168.200.0/24
/home/pvdata/mysql     192.168.200.0/24
```

可以看到创建的两个目录被共享了,NFS 配置完毕。

(2)创建 PV

在创建 PV 之前,首先创建一个工作目录,命令如下:

```
[root@master ~]# mkdir wordpress
[root@master ~]# cd wordpress/
[root@master wordpress]#
```

在 WordPress 目录下创建 mysql-persistent-storage.yaml 文件,该文件用来创建数据库所使用的 PV,文件内容如下:

```
[root@master wordpress]# cat mysql-persistent-storage.yaml
apiVersion: v1
kind: PersistentVolume
metadata:
  name: mysql-persistent-storage
spec:
  capacity:
    storage: 2Gi
  accessModes: ["ReadWriteMany","ReadWriteOnce"]
  persistentVolumeReclaimPolicy: Recycle
  nfs:
    path: /home/pvdata/mysql/
    server: 192.168.200.19
```

在 WordPress 目录下继续创建 wordpress-persistent-storage.yaml 文件,该文件用来创建 WordPress 所使用的 PV,文件内容如下:

```
[root@master wordpress]# cat wordpress-persistent-storage.yaml
apiVersion: v1
kind: PersistentVolume
metadata:
  name: wordpress-persistent-storage
spec:
  capacity:
    storage: 2Gi
  accessModes: ["ReadWriteMany","ReadWriteOnce"]
  persistentVolumeReclaimPolicy: Recycle
  nfs:
    path: /home/pvdata/wordpress/
    server: 192.168.200.19
```

编辑好 Yaml 文件后,创建 PV 卷,命令如下:

```
[root@master wordpress]# kubectl apply -f mysql-persistent-storage.yaml
persistentvolume/mysql-persistent-storage created
[root@master wordpress]# kubectl apply -f wordpress-persistent-storage.yaml
persistentvolume/wordpress-persistent-storage created
```

创建完 PV 后，查看 PV，命令如下：

```
[root@master wordpress]# kubectl get pv
NAME                          CAPACITY    ACCESS MODES    RECLAIM POLICY    STATUS      CLAIM
STORAGECLASS    REASON    AGE
mysql-persistent-storage         2Gi       RWO,RWX    Recycle      Available   11s
wordpress-persistent-storage     2Gi       RWO,RWX    Recycle      Available   5s
```

可以看到两个 PV 都处于 Available 的状态，PV 创建成功。

（3）设置密码生成器

在/root/wordpress 目录下创建 kustomization.yaml 文件，并在文件中添加一个 Secret 生成器，具体命令如下：

```
[root@master wordpress]# cat kustomization.yaml
secretGenerator:
- name: mysql-pass
  literals:
  - password=123456ok
```

（4）编辑 Yaml 文件

首先在/root/wordpress 目录下创建一个 mysql-deployment.yaml 文件，用来部署数据库服务，MySQL 容器将 PersistentVolume 挂载在/var/lib/mysql。MYSQL_ROOT_PASSWORD 环境变量设置来自 Secret 的数据库密码，mysql-deployment.yaml 中的具体内容如下所示：

原生 Kubernetes 容器云平台应用部署（2）

```
[root@master wordpress]# cat mysql-deployment.yaml
apiVersion: v1
kind: Service
metadata:
  name: wordpress-mysql
  labels:
    app: wordpress
spec:
  ports:
    - port: 3306
  selector:
    app: wordpress
    tier: mysql
  clusterIP: None
---
apiVersion: v1
```

```yaml
kind: PersistentVolumeClaim
metadata:
  name: mysql-pv-claim
  labels:
    app: wordpress
spec:
  accessModes: ["ReadWriteOnce"]
  resources:
    requests:
      storage: 2Gi
---
apiVersion: apps/v1 # for versions before 1.9.0 use apps/v1beta2
kind: Deployment
metadata:
  name: wordpress-mysql
  labels:
    app: wordpress
spec:
  selector:
    matchLabels:
      app: wordpress
      tier: mysql
  strategy:
    type: Recreate
  template:
    metadata:
      labels:
        app: wordpress
        tier: mysql
    spec:
      containers:
      - image: mysql:5.6
        name: mysql
        env:
        - name: MYSQL_ROOT_PASSWORD
          valueFrom:
            secretKeyRef:
              name: mysql-pass
              key: password
        ports:
        - containerPort: 3306
          name: mysql
        volumeMounts:
        - name: mysql-persistent-storage
          mountPath: /var/lib/mysql
```

```
      volumes:
      - name: mysql-persistent-storage
        persistentVolumeClaim:
          claimName: mysql-pv-claim
```

编辑好 mysql-deployment.yaml 之后,数据库的部署就告一段落了。接下来还需要创建一个 wordpress-deployment.yaml 用来部署 WordPress 应用,WordPress 容器网站数据文件位于 /var/www/html。WORDPRESS_DB_HOST 环境变量集上定义了 MySQL 的 Service 名称,WordPress 将通过 Service 访问数据库。WORDPRESS_DB_PASSWORD 密码为 Secret 生成的数据库密码。具体文件内容如下:

```
[root@master wordpress]# cat wordpress-deployment.yaml
apiVersion: v1
kind: Service
metadata:
  name: wordpress
  labels:
    app: wordpress
spec:
  ports:
    - port: 80
      nodePort: 31000
  selector:
    app: wordpress
    tier: frontend
  type: NodePort
---
apiVersion: v1
kind: PersistentVolumeClaim
metadata:
  name: wp-pv-claim
  labels:
    app: wordpress
spec:
  accessModes: ["ReadWriteOnce"]
  resources:
    requests:
      storage: 2Gi
---
apiVersion: apps/v1 # for versions before 1.9.0 use apps/v1beta2
kind: Deployment
metadata:
  name: wordpress
  labels:
```

```
    app: wordpress
spec:
  selector:
    matchLabels:
      app: wordpress
      tier: frontend
  strategy:
    type: Recreate
  template:
    metadata:
      labels:
        app: wordpress
        tier: frontend
    spec:
      containers:
      - image: wordpress:4.8-apache
        name: wordpress
        env:
        - name: WORDPRESS_DB_HOST
          value: wordpress-mysql
        - name: WORDPRESS_DB_PASSWORD
          valueFrom:
            secretKeyRef:
              name: mysql-pass
              key: password
        ports:
        - containerPort: 80
          name: wordpress
        volumeMounts:
        - name: wordpress-persistent-storage
          mountPath: /var/www/html
      volumes:
      - name: wordpress-persistent-storage
        persistentVolumeClaim:
          claimName: wp-pv-claim
```

可以注意到,此次部署的 WordPress 版本为 WordPress:4.8-apache,该镜像在配套的软件包中有提供,可自行上载到 master 节点。编辑完之后,保存退出,将以上两个文件补充到 kustomization.yaml 文件中,命令如下:

```
[root@master wordpress]# vi kustomization.yaml
[root@master wordpress]# cat kustomization.yaml
secretGenerator:
- name: mysql-pass
```

```
    literals:
    - password=123456ok
#在配置文件中添加如下代码
resources:
    - mysql-deployment.yaml
    - wordpress-deployment.yaml
```

kustomization.yaml 包含用于部署 WordPress 网站的所有资源以及 MySQL 数据库。

（5）部署应用

在/root/wordpress 目录下，通过如下方式进行部署，命令如下：

```
[root@master wordpress]# kubectl apply -k ./
secret/mysql-pass-ccht6ddg2m created
service/wordpress-mysql created
service/wordpress created
deployment.apps/wordpress-mysql created
deployment.apps/wordpress created
persistentvolumeclaim/mysql-pv-claim created
persistentvolumeclaim/wp-pv-claim created
```

可以看到各类资源都被创建了，包括 Service、Deployment、PVC 等。查看 Secret 密码是否存在，命令如下：

```
[root@master wordpress]# kubectl get secret
NAME                    TYPE        DATA    AGE
mysql-pass-ccht6ddg2m   Opaque      1       5h2m
```

可以看到密码被成功创建，然后验证 PVC 是否被成功挂载了，命令如下：

```
[rootgmaster wordpress]#     kubectl get pvc              AGE
NAME            STATUS   VOLUME                     CAPACITY  ACCESS MODES
STORAGECLASS    AGE
mysql-pv-claim  Bound    mysql-persistent-storage    2Gi      RWO, RWX    3h1m
wp- pv-claim    Bound    wordpress-persistent-storage 2Gi     RwO, Rwx    3h1m
```

可以看到创建的 PVC 被挂载了。最后验证 Service 是否被成功创建，命令如下：

```
[rootamaster wordpress]# kubectl get svc
NAME              TYPE        CLUSTER-IP      EXTERNAL-IP   PORT(S)         AGE
kubernetes        clusterIP   10.96.0.1       <none>        443/TCP         102d
wordpress         NodePort    10.101.227.212  <none>        80: 31000/TCP   3h2m
wordpress-mysql   clusterIP   None            <none>        3306/TCP        3h2m
```

也可以看到 Service 被成功创建。

（6）验证应用

通过浏览器访问 http://192.168.200.19:31000，如图 8-9 所示。

根据提示选择语言，并单击"Continue"按钮，进入配置信息界面，如图 8-10 所示。

图 8-9　WordPress 安装页面

图 8-10　WordPress 配置信息界面

根据要求填写相关信息，然后单击下方的"Install WordPress"按钮进行安装，然后就可以登录了。使用自己设置的用户名和密码进行登录，如图 8-11 所示。

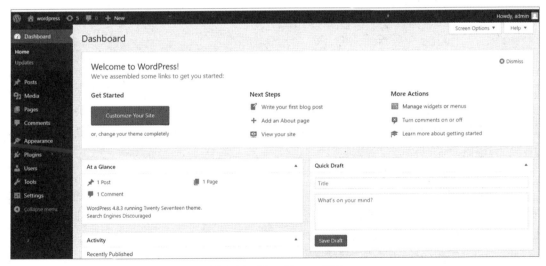
图 8-11　WordPress 首页

登录成功，则表示部署 WordPress 应用成功。通过上述案例，掌握了如何创建 PV 卷并创建 PVC 挂载使用，也对 Yaml 的编写有了更进一步的了解。感兴趣的读者可以深入了解 Kubernetes 中其他组件的使用。

4. Kubernetes 运维

通过上述内容的学习，相信读者已经对 Kubernetes 服务有了很深刻的认识，接下来介绍在使用 Kubernetes 服务时，经常会使用到的一些运维操作。

（1）Node 的隔离与恢复

在硬件升级、硬件维护等情况下，需要将某些 Node 节点隔离。使用 kubectl cordon <node_name>命令可禁止 Pod 调度到该节点上，但在其上运行的 Pod 并不会自动停止，管理员需要手动停止在该 Node 节点上运行的 Pod。此处使用的实操环境是一台 PC，尝试将 master 节点隔离（master 也是一个 Node 节点），命令如下：

```
[root@master ~]# kubectl cordon master
node/master cordoned
```

查看 Node 节点的状态，命令如下：

```
[root@master ~]# kubectl get nodes
NAME     STATUS                     ROLES    AGE    VERSION
master   Ready,SchedulingDisabled   master   102d   v1.18.1
```

可以观察到在 Node 节点的状态中增加了一项 SchedulingDisabled，对于后续创建的 Pod，系统将不会再向该 Node 节点进行调度。

可以通过 kubectl uncordon 命令完成对 Node 节点的恢复，命令如下：

```
[root@master ~]# kubectl uncordon master
node/master uncordoned
```

再次查看 Node 节点信息，命令如下：

```
[root@master ~]# kubectl get nodes
NAME     STATUS   ROLES    AGE    VERSION
master   Ready    master   102d   v1.18.1
```

可以看到 Node 节点已恢复调度，STATUS 又变成了 Ready，即允许 Pod 调度到该节点上。

通过 kubectl drain <node_name>命令可实现对 Node 节点的驱逐，该命令会删除该节点上的所有 Pod（DaemonSet 除外），并在其他 Node 节点上重新启动它们。此处就不再演示了，感兴趣的读者可以自行尝试。

（2）Pod 动态扩容和缩放

在实际生产系统中，经常会遇到某个服务需要扩容的场景，也可能会遇到由于资源紧张或者工作负载降低而需要减少服务实例数量的场景。此时可以利用 kubectl scale deployment 命令来完成这些任务。

以 Nginx Deployment 为例，创建一个 Nginx 的 Deployment，副本数为 1，命令如下（在做该案例时，为了避免 Kubernetes 实战中创建的资源受影响，可自行删除）：

```
[root@master ~]# kubectl create deployment nginx --image=nginx:latest
deployment.apps/nginx created
```

查看 Pod 和 Deployment，命令如下：

```
[root@master ~]# kubectl get pods
NAME                    READY   STATUS    RESTARTS   AGE
nginx-674ff86d-n6jqq    1/1     Running   0          2m1s
```

```
[root@master ~]# kubectl get deployment
NAME       READY   UP-TO-DATE   AVAILABLE   AGE
nginx      1/1     1            1           2m7s
```

通过使用 kubectl scale 命令将 Nginx Deployment 控制的 Pod 副本数量从初始的 1 更新为 5，命令如下：

```
[root@master ~]# kubectl scale deployment nginx --replicas=5
deployment.apps/nginx scaled
```

更新副本后，继续查看 Pod 数量，命令如下：

```
[root@master ~]# kubectl get pods
NAME                      READY   STATUS    RESTARTS   AGE
nginx-674ff86d-7xgct      1/1     Running   0          66s
nginx-674ff86d-9mzvx      1/1     Running   0          66s
nginx-674ff86d-dw6mb      1/1     Running   0          66s
nginx-674ff86d-n6jqq      1/1     Running   0          4m33s
nginx-674ff86d-ztm8j      1/1     Running   0          66s
```

可以看到此时 Nginx 的 Pod 数量为 5，调整副本数成功。如果需要将副本的数量调小，可以将--replicas 设置为比当前 Pod 副本数量更小的数字，系统将会"杀掉"一些运行中的 Pod，即可实现应用集群缩容。例如将 Nginx 的 Pod 副本数调整为 3，命令如下：

```
[root@master ~]# kubectl scale deployment nginx --replicas=3
deployment.apps/nginx scaled
```

查看 Pod 数量，命令如下：

```
[root@master ~]# kubectl get pods
NAME                      READY   STATUS    RESTARTS   AGE
nginx-674ff86d-9mzvx      1/1     Running   0          4m56s
nginx-674ff86d-dw6mb      1/1     Running   0          4m56s
nginx-674ff86d-n6jqq      1/1     Running   0          8m23s
```

可以看到 Pod 副本数已经减少为 3，验证 Pod 动态扩容和缩放实验成功。

（3）将 Pod 调度到指定的 Node 节点

Kubernetes 的 Scheduler 服务（Kube-Scheduler 进程）负责实现 Pod 的调度，整个调度过程通过执行一系列复杂的算法最终为每个 Pod 计算出一个最佳的目标节点，这一过程是自动完成的，用户无法知道 Pod 最终会被调度到哪个节点上。有时可能需要将 Pod 调度到一个指定的 Node 节点上，此时，可以通过 Node 节点的标签（Label）和 Pod 的 nodeSelector 属性相匹配，来达到上述目的。

Label（标签）作为用户可灵活定义的对象属性，在已创建的对象上，仍然可以随时通过 kubectl label 命令对其进行增加、修改、删除等操作。使用 kubectl label 给 Node 节点打标签的用法如下：

```
# kubectl label nodes <node-name> <label-key>=<label-value>
```

因为该案例起码需要两个 Node 节点，所以还需申请一台虚拟机，配置为 Kubernetes 集群的 Node 节点。根据 Kubernetes 部署的内容，自行安装 Kubernetes 的 Node 节点，然后进行下述操作。

为 Node 节点打上一个 project=gcxt 的标签，命令如下：

```
[root@master ~]# kubectl label nodes node project=gcxt
node/node labeled
```

如果想删除 Label，只需要在命令行的最后指定 Label 的 key 名，并加一个减号即可。

```
[root@master ~]# kubectl label node node project-
node/node labeled
```

在 Pod 中加入 nodeSelector 定义，示例如下：

```
[root@master ~]# cat nginx.yaml
apiVersion: v1
kind: ReplicationController
metadata:
  name: nginx-gcxt
  labels:
    name: nginx-gcxt
spec:
  replicas: 1
  selector:
    name: nginx-gcxt
  template:
    metadata:
      labels:
        name: nginx-gcxt
    spec:
      containers:
      - name: nginx-gcxt
        image: nginx: latest
        - containerPort: 80
      nodeSelector:
        project: gcxt
```

运行 kubectl apply -f 命令创建 Pod，Scheduler 就会将该 Pod 调度到拥有 project=gcxt 标签的 Node 节点上去。

```
[root@master ~]# kubectl apply -f nginx.yaml
replicationcontroller/nginx-gcxt created
```

查看 Pod 的详细信息，命令如下：

```
[root@master ~]# kubectl get pods -owide
NAME            READY   STATUS    RESTARTS   AGE   IP         NODE
```

```
NOMINATED NODE   READINESS GATES
  nginx-gcxt-hdt5x   1/1       Running   0           14s      10.24.9.2    node
<none>           <none>
```

可以看到,Pod 已被成功调度到指定的 Node 节点。这种基于 Label 标签的调度方式灵活性很高,比如,可以把一组 Node 节点分别贴上"开发环境""测试环境""生产环境"这 3 组标签中的一种,此时一个 Kubernetes 集群就承载了 3 个环境,这将大大提高开发效率。

注意:如果指定了 Pod 的 nodeSelector 条件,且集群中不存在包含相应标签的 Node 节点,那么即使还有其他可供调度的 Node 节点,这个 Pod 也最终会调度失败。

(4)应用滚动升级

当集群中的某个服务需要升级时,需要停止目前与该服务相关的所有 Pod,然后重新拉取镜像并启动。如果集群规模比较大,这个工作就变成了一个挑战。如果采取先全部停止,然后逐步升级的方式,就会导致较长时间的服务不可用。Kubernetes 提供了 rolling-update(滚动升级)功能来解决上述问题。

滚动升级通过执行 kubectl rolling-update 命令一键完成,该命令创建了一个新的 Deployment,然后自动控制旧的 Deployment 中的 Pod 副本数量逐渐减少到 0,同时新的 Deployment 中的 Pod 副本数量从 0 逐步增加到目标值,最终实现 Pod 的升级。

注意:系统要求新的 Deployment 需要与旧的 Deployment 在相同的命名空间(Namespace)内,即不能把别人的资产偷偷转移到自家名下。

下面的示例在第一次部署时使用 httpd: 2.2.31,然后使用滚动升级更新到 httpd: 2.2.32。

首先定义 Yaml 文件,在/root 目录下创建 httpd.yaml 文件,文件内容如下:

```
[root@master ~]# vi httpd.yaml
[root@master ~]# cat httpd.yaml
apiVersion: apps/v1
kind: Deployment
metadata:
  name: httpd
spec:
  selector:
    matchLabels:
      app: httpd
  replicas: 3
  template:
    metadata:
      labels:
        app: httpd
    spec:
      containers:
      - name: httpd
        image: httpd:2.2.31
        ports:
        - containerPort: 80
```

创建 httpd 的 Deployment，命令如下：

```
[root@master ~]# kubectl apply -f httpd.yaml
deployment.apps/httpd created
```

查看 Pod 和 Deployment 信息，命令如下：

```
[root@master ~]# kubectl get pod
NAME                        READY   STATUS    RESTARTS   AGE
httpd-745bddc8fb-7h9cr      1/1     Running   0          9m22s
httpd-745bddc8fb-h2s2p      1/1     Running   0          9m22s
httpd-745bddc8fb-wtx8r      1/1     Running   0          9m22s
[root@master ~]# kubectl get deployment -o wide
NAME    READY   UP-TO-DATE   AVAILABLE   AGE   CONTAINERS   IMAGES         SELECTOR
httpd   3/3     3            3           16m   httpd        httpd:2.2.31   app=httpd
```

可以看到 images 的版本是 2.2.31，修改 httpd.yaml 文件，把配置文件中的 httpd:2.2.31 改为 httpd:2.2.32，命令如下：

```
[root@master ~]# vi httpd.yaml
[root@master ~]# cat httpd.yaml
apiVersion: apps/v1
kind: Deployment
metadata:
  name: httpd
spec:
  selector:
    matchLabels:
      app: httpd
  replicas: 3
  template:
    metadata:
      labels:
        app: httpd
    spec:
      containers:
      - name: httpd
        image: httpd:2.2.32
        ports:
        - containerPort: 80
```

再次启动 Deployment，命令如下：

```
[root@master ~]# kubectl apply -f httpd.yaml
deployment.apps/httpd configured
```

再次查看 Deployment 的信息，命令如下：

```
[root@master ~]# kubectl get deployment -o wide
NAME    READY   UP-TO-DATE   AVAILABLE   AGE   CONTAINERS   IMAGES        SELECTOR
httpd   3/3     3            3           16m   httpd        httpd:2.2.32  app=httpd
```

可以看到当前的 httpd 版本为 2.2.32，查看 Deployment 的详细信息，命令如下：

```
[root@master ~]# kubectl describe deployment httpd
Name:                   httpd
Namespace:              default
CreationTimestamp:      Mon, 27 Sep 2021 11:20:28 -0400
Labels:                 <none>
Annotations:            deployment.kubernetes.io/revision: 2
Selector:               app=httpd
Replicas:               3 desired | 3 updated | 3 total | 3 available | 0 unavailable
StrategyType:           RollingUpdate
MinReadySeconds:        0
RollingUpdateStrategy:  25% max unavailable, 25% max surge
Pod Template:
  Labels:  app=httpd
  Containers:
   httpd:
    Image:        httpd:2.2.32
    Port:         80/TCP
    Host Port:    0/TCP
    Environment:  <none>
    Mounts:       <none>
  Volumes:        <none>
Conditions:
  Type           Status  Reason
  ----           ------  ------
  Available      True    MinimumReplicasAvailable
  Progressing    True    NewReplicaSetAvailable
OldReplicaSets:  <none>
NewReplicaSet:   httpd-55d897fbfc (3/3 replicas created)
Events:
  Type    Reason             Age   From                   Message
  ----    ------             ----  ----                   -------
  Normal  ScalingReplicaSet  10m   deployment-controller  Scaled up replica set httpd-745bddc8fb to 3
  Normal  ScalingReplicaSet  18s   deployment-controller  Scaled up replica set httpd-55d897fbfc to 1
  Normal  ScalingReplicaSet  16s   deployment-controller  Scaled down
```

```
replica set httpd-745bddc8fb to 2
    Normal  ScalingReplicaSet  16s  deployment-controller  Scaled up replica set httpd-55d897fbfc to 2
    Normal  ScalingReplicaSet  14s  deployment-controller  Scaled down replica set httpd-745bddc8fb to 1
    Normal  ScalingReplicaSet  14s  deployment-controller  Scaled up replica set httpd-55d897fbfc to 3
    Normal  ScalingReplicaSet  13s  deployment-controller  Scaled down replica set httpd-745bddc8fb to 0
```

上面的日志信息描述了滚动升级的过程：

① 启动一个新版 Pod。

② 把旧版 Pod 数量降为 2。

③ 再启动一个新版 Pod，数量变为 2。

④ 把旧版 Pod 数量降为 1。

⑤ 再启动一个新版 Pod，数量变为 3。

⑥ 把旧版 Pod 数量降为 0。

这就是滚动的意思，始终保持副本数量为 3，控制新旧 Pod 的交替，实现了无缝升级。

归纳总结

通过本单元内容的学习，掌握了 Kubernetes 服务的概念、特点、应用场景与发展历程。同时，通过 Kubernetes 的部署、使用、实战、运维等实操案例，让读者对 Kubernetes 服务的部署使用有了更深刻的理解。关于更多 Kubernetes 服务的使用与实践，读者可以自行查找资料学习。

课后练习

一、判断题

1. 在 K8S 的核心组件中负责维护容器生命周期的组件是 Kubelet。（   ）

2. 在 K8S 的核心组件中负责资源调度的组件是 Scheduler。（   ）

二、单项选择题

1. 下面（   ）是 Kubernetes 可操作的最小对象。

A. Container　　　　B. Pod　　　　　　C. Image　　　　　D. Volume

2. 下面（   ）是 Kubernetes 的核心组件。

A. Neutron　　　　　B. Scheduler　　　　C. Daemon　　　　D. Cgroup

三、多项选择题

1. 下面关于 Kubernetes 的说法中，正确的是（   ）。

A. Kubernetes 是开源的容器集群管理系统

B. Kubenetes 中，所有的容器均在 Pod 中运行，一个 Pod 只能承载一个容器

C. Kubernetes 的核心特点是能够自主地管理容器

D. Kubernetes 可操作的最小对象是 Pod

2. 下面关于 Kubernetes 架构的说法中，不正确的是（　　）。

A. Kubernetes 集群包含所有节点代理 Kubelet 和 Master 组件

B. Kubernetes 节点有运行应用容器必备的服务，而这些都受 Kubelet 的控制

C. Kubernetes 集群只需要管理节点运行 Docker

D. Kubernetes 集群中有管理节点与工作节点两种类型

技能训练

1. 安装 Docker 服务并搭建 Kubernetes 集群服务，查询集群节点的健康状态。
2. 在集群节点进行运维操作，完成对 HTTP 服务应用的回滚操作。

单元 9 基于 K8S 的 CI/CD 服务

学习目标

通过本单元的学习，要求了解什么是持续集成/持续交付（CI/CD），CI/CD 在日常工作中起到了什么作用、解决了哪些问题，构建 CI/CD 服务需要用到哪些工具，了解 CI/CD 的优点等。本单元培养读者掌握 CI/CD 环境的构建、CI/CD 环境的配置与触发等技能；也培养读者的自我学习能力和逻辑思维能力。

9.1 CI/CD 服务

9.1.1 CI/CD 服务介绍

持续集成（CI）和持续交付（CD）都体现了如今快节奏市场中的文化和发展原则，旨在缩短开发周期、提高软件交付效率以及实现全流程的自动化。同时，两者都有着共同的目标：让软件开发更少地依赖于手动执行的任务，在此基础上使得软件的发布更加频繁、更加安全可靠。由于有着相同的目标，因此持续集成和持续交付并非相互排斥，只是它们的应用范围有所不同。

1. 持续集成简介

在了解持续集成之前，先了解什么叫集成，集成的英文翻译为 Continuous Integration（简称 CI）。

持续集成是一种软件开发实践，即团队开发成员经常集成用户的工作，通常每个成员每天至少集成一次，也就意味着每天可能会发生多次集成。每次集成都通过自动化的构建（包括编译、发布、自动化测试）来验证，从而尽快地发现集成错误。许多团队发现这个过程可以大大减少集成的问题，让团队能够更快地开发软件。

如果没有持续集成，则会发生以下情况：

① 项目做模块集成的时候，发现很多接口都不通，从而浪费大量时间。
② 需要人手动去编译打包最新的代码，构建过程不透明。
③ 发布代码、上线等基本靠手工，脚本太多。

使用持续集成可以：

① 维护一个单一的代码库。
② 使构建自动化。

③ 执行测试是构建的一部分。
④ 集成日志及历史记录。
⑤ 使用统一的依赖包管理库。
⑥ 每天至少集成一次。

2. 持续交付介绍

持续集成包含了构建与自动化测试的阶段，而持续交付（Continuous Delivery，CD）在持续集成的基础上，将集成后的代码部署到更贴近真实运行环境的"类生产环境"之中。持续交付优先于整个产品生命周期的软件部署，建立在高水平自动化的持续集成之上。此外，持续交付同样遵循一个小型的构建周期，可以将一小批代码推送到多个环境：开发、测试或生产。

在此过程中，它结合了持续集成和持续部署的实践（即让主干始终处于可部署状态）。当提交的代码完成以上步骤后，可以在"部署流水线—流水线管理"中创建对应的 CD 流水线将持续集成后产生的应用版本自动部署到对应的环境中去。此外，对于部署到正式环境中的代码，可以在流水线中添加一个人工卡点任务，只有通过人工审核后，才能执行后续的自动部署任务。

9.1.2 CI/CD 的工作流程与使用场景

1. CI/CD 的工作流程

持续集成（CI），是软件开发周期的一种实践，把代码仓库（GitLab 或者 GitHub）、构建工具（如 Jenkins）和测试工具（如 SonarQube）集成在一起，频繁地将代码合并到主干然后自动进行构建和测试。简单来说，持续集成就是一个监控并控制系统中代码变化的工具，当发生变化时可以自动编译、测试以及执行后续的自定义动作。持续集成的工作流程如图 9-1 所示。

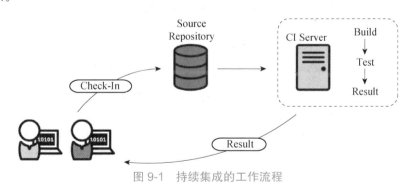

图 9-1 持续集成的工作流程

开发人员先把代码放到 GitHub，然后 Jenkins 从 GitHub 获取代码进行构建、测试、生成结果，再返回给客户端。

而持续交付 CD 则在 CI 的基础进行了扩展，CI 环节完成了软件构建和测试工作，并形成了新的版本，那么接下来就要进行交付，而这里的交付并不是交付到生产环境，而是类生产环境（Staging），用户可以理解为灰度环境或者预发环境，进而接受部分真实流量的测试。如果没有问题，则通过手动的方式部署到生产环境。持续交付的工作流程如图9-2所示。

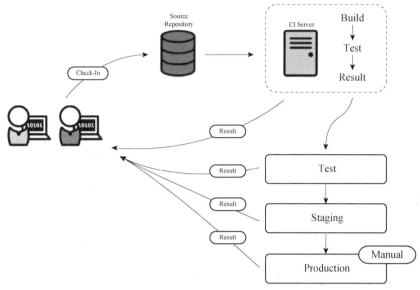

图 9-2　持续交付的工作流程

持续集成和持续交付的目的是减少代码改动到投入生产过程的所需时间，提早发现风险、减少测试时长、减少运维的人工干预，整体上是一个提效的过程。

2. CI/CD 的使用场景

持续集成可确保代码库中始终保存最新的代码，同时可以快速集成来自多个开发人员的代码，并确保这些代码可在多个环境中协同工作。它通常有助于减少错误并通过自动化流程来减少手动任务。CI 可以实现代码的自动构建与测试，减少开发中的 Bug。因此，CI 适用于那些过度依赖手动任务和复杂构建过程的企业。

持续交付适用于需要缩短开发周期，更快地为目标用户提供软件的企业。CD 降低了部署新软件或升级已有软件的难度，且实现了全流程的自动化，因此团队中的工作人员无须手动执行复杂烦琐的任务，从而加快反馈速度，来确保增加的功能真正地满足用户的需求。

总而言之，CI 和 CD 是相互补充的。CI 的统一代码库和自动化测试的方法可用于支持 CD 中更大规模的自动化和更频繁的部署。因此将 CI 和 CD 结合到开发与交付的流程中，会使 IT 团队更加敏捷更加快速地开发。

9.1.3　CI/CD 的优点

使用 CI/CD 的优点如下。

1. 软件交付更快

在产品的开发过程中借助自动化，可以更快地交付产品，并且可以获得最终用户的反馈，因为需要完成的手动任务较少，可以更频繁地发布较小的变更到生产环境。

2. 提高可见度

使用 CI/CD，其测试和部署过程是透明的。任何问题都可以被立即看到，并且可以快速找到原因，从而减少了排查问题的时间。由于漏洞易于跟踪，所以能够快速找到解决该问题的最佳人选，最终，增加了系统正常运行时间。

3. 错误减少

现在软件的功能、项目和应用程序都很复杂，错误也越来越复杂。持续集成消除了其中的一些复杂性，减少了问题所在的领域，并增加了成功的可能性。自动化可以减少人为错误。

持续集成的两个主要原则是频繁运行代码以及自动执行软件交付生命周期的构建和测试部分。每天多次运行代码集成意味着可以更快地发现错误，然后可以更快地纠正错误。

开发人员的手动任务更少，这意味着人为错误机会更少，从而，运维部门会收到高质量的代码，需要解决的问题也会变少；客服也不会收到那么多恼人的客户投诉邮件或投诉电话；每个人的工作效率都得到改善。

4. 提高人员效率

如果将可重复和可预测的任务移交给自动化，那么开发人员就能够专注于做开发工作。将烦琐、重复的任务留给计算机，让开发人员完成更有意义、更有创意的工作。

5. 提高满意度

使用 CI/CD，可以更快、更频繁地发布项目，同时错误也会变少。这可以增加开发人员与其他业务部门之间的信任。按时完成任务、获得可靠的结果可以使最终用户更加满意。

9.2 基于 K8S 构建持续集成

1. 环境准备

（1）规划节点

持续集成案例实验节点如表 9-1 所示。

表 9-1 持续集成案例实验节点

IP 地址	主机名	节点
192.168.200.20	master	Kubernetes 实验节点
192.168.200.21	node	Kubernetes 实验节点

（2）基础准备

准备好 VMware Workstation 工具和 CentOS7.5 版本虚拟机，使用 CentOS7.5 虚拟机快照，克隆两台虚拟机，云主机配置自定义，推荐配置 2vCPU、6GB 内存、60GB 硬盘，网络使用 NAT 模式，master 节点 IP 设置为 192.168.200.20，node 节点 IP 设置为 192.168.200.21，推荐设置主机密码为 000000。自行将虚拟机配置好 IP、安装 Docker 服务，关闭 SELinux

和防火墙服务（安装 Docker 服务的实验步骤请参照 1.2 节，安装 K8S 集群的实验步骤参照 8.2 节）。

2. CI/CD 环境构建

（1）基础环境准备

查看已经安装好的 Kubernetes 集群状态，命令如下：

```
[root@master ~]# kubectl get nodes
NAME     STATUS   ROLES    AGE     VERSION
master   Ready    master   37m     v1.18.1
node     Ready    <none>   3m59s   v1.18.1
```

将提供的离线包 CICD_Offline.tar 上传至 master 节点/root 目录下，并解压到目标目录，命令如下：

```
[root@master ~]# tar -zxvf CICD_Offline.tar -C /opt/
```

解压完毕之后，进入/opt 目录，将 Jenkins 镜像导入，命令如下：

```
[root@master ~]# cd /opt/
[root@master opt]# docker load -i jenkins.tar
```

导入之后，基础环境准备完毕。

（2）安装 Jenkins

在 master 节点的/root 目录下创建 Jenkins 工作目录，编写 Jenkins 的 docker-compose.yaml 编排文件，文件内容如下：

```
[root@master ~]# mkdir jenkins
[root@master ~]# cd jenkins
[root@master jenkins]# vi docker-compose.yaml
[root@master jenkins]# cat docker-compose.yaml
version: '3.1'
services:
  jenkins:
    image: 'jenkins/jenkins:2.262-centos'
    volumes:
      - /home/jenkins_home:/var/jenkins_home
      - /var/run/docker.sock:/var/run/docker.sock
      - /usr/bin/docker:/usr/bin/docker
      - /usr/bin/kubectl:/usr/local/bin/kubectl
      - /root/.kube:/root/.kube
    ports:
      - "8080:8080"
    expose:
      - "8080"
```

```
      - "50000"
    privileged: true
    user: root
    restart: always
    container_name: jenkins
```

启动 Jenkins 服务，命令如下：

```
[root@master jenkins]# docker-compose -f docker-compose.yaml up -d
Creating network "root_default" with the default driver
Creating jenkins ... done
[root@master jenkins]# docker-compose up -d
Creating network "jenkins_default" with the default driver
Creating jenkins ... done
[root@master jenkins]# docker-compose ps
 Name         Command              State              Ports
-------------------------------------------------------------------------------
 jenkins    /sbin/tini -- /usr/local/b ...    Up       50000/tcp,
0.0.0.0:8080->8080/tcp
```

安装 Jenkins 的插件，命令如下：

```
[root@master jenkins]# cp -rfv /opt/plugins/* /home/jenkins_home/plugins/
[root@master jenkins]# docker restart jenkins
jenkins
```

启动 Jenkins 服务完成后，可以在 Web 端通过 http://IP:8080 访问 Jenkins，如图 9-3 所示。

图 9-3　Jenkins 登录界面

登录 Jenkins 需要密码，回到 master 节点查看密码，命令如下：

```
[root@master ~]# docker exec jenkins cat /var/jenkins_home/secrets/initialAdminPassword
54670064056e42a2a47485a5e3356e57
```

输入查询到的密码并单击"继续",进入 Jenkins 首页,如图 9-4 所示。

图 9-4　Jenkins 首页

依次单击左侧导航栏的"系统管理"→"管理用户"按钮进入用户列表界面,如图 9-5 所示。

图 9-5　用户列表界面

单击左侧的"新建用户",输入用户信息,进行创建用户操作,如图 9-6 所示。

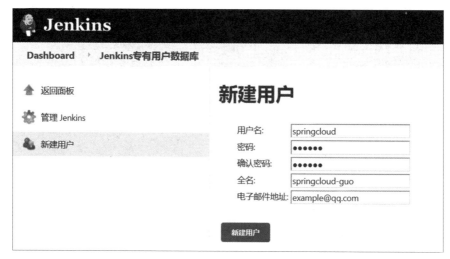

图 9-6　新建用户界面

单击"新建用户"按钮，完成用户的创建，如图 9-7 所示。

图 9-7 新建用户

新建用户完毕后，退出 admin 用户登录，使用新创建的 springcloud 用户登录 Jenkins。

单击"系统配置"按钮进入系统配置界面，在"Resource root URL"处配置 Jenkins URL，如图 9-8 所示。

图 9-8 URL 配置界面

配置完成后单击"保存"按钮。

（3）部署 GitLab

GitLab 是一个开源的版本管理系统实现的自托管的 Git 项目仓库，可通过 Web 界面访问公开的或者私人项目。与 GitHub 类似，GitLab 能够浏览源代码、管理缺陷和注释，可以管理团队对仓库的访问，它非常易于浏览提交过的版本并提供了一个文件历史库，团队成员可以利用内置的简单聊天程序（Wall）进行交流。GitLab 还提供了一个代码片段收集功能，可以轻松实现代码复用，便于日后有需要的时候进行查找。本项目中 GitLab 与 Harbor 共用一台服务器。

① 启动 GitLab。在 master 节点的/root 目录下创建 gitlab 工作目录，然后编写 GitLab 编排文件，命令如下：

```
[root@master ~]# mkdir gitlab
[root@master ~]# cd gitlab/
[root@master gitlab]# vi docker-compose.yaml
version: '3'
services:
  gitlab:
    image: 'gitlab/gitlab-ce:12.9.2-ce.0'
    container_name: gitlab
```

```
    restart: always
    hostname: '10.24.2.156'
    privileged: true
    environment:
      TZ: 'Asia/Shanghai'
    ports:
      - '81:80'
      - '443:443'
      - '1022:22'
    volumes:
      - /srv/gitlab/config:/etc/gitlab
      - /srv/gitlab/gitlab/logs:/var/log/gitlab
      - /srv/gitlab/gitlab/data:/var/opt/gitlab
```

编写完 Yaml 文件后，启动 GitLab，命令如下：

```
[root@master gitlab]# docker-compose up -d
Creating network "gitlab_default" with the default driver
Creating gitlab ... done
[root@master gitlab]# docker-compose ps
 Name          Command              State                      Ports
---------------------------------------------------------------------------------------
 gitlab    /assets/wrapper      Up (healthy)   0.0.0.0:1022->22/tcp, 0.0.0.0:
443->443/tcp, 0.0.0.0:81->80/tcp
```

GitLab 启动较慢，可以通过 docker logs 命令查看启动状态。启动完成后，在 Web 端访问 GitLab（http://IP:81），如图 9-9 所示。

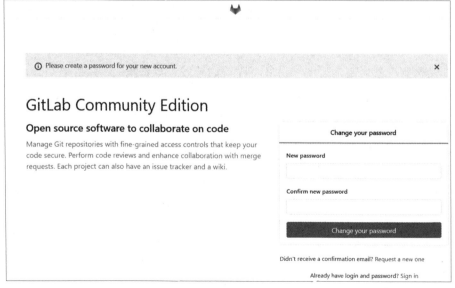

图 9-9　GitLab 登录界面

设置 root 用户密码，并使用 root 用户登录 GitLab，登录之后的首页如图 9-10 所示。

图 9-10　Gitlab 首页

② 创建项目。单击"Create a project"按钮，创建项目"springcloud"，可见等级选择"Public"，如图 9-11 所示。

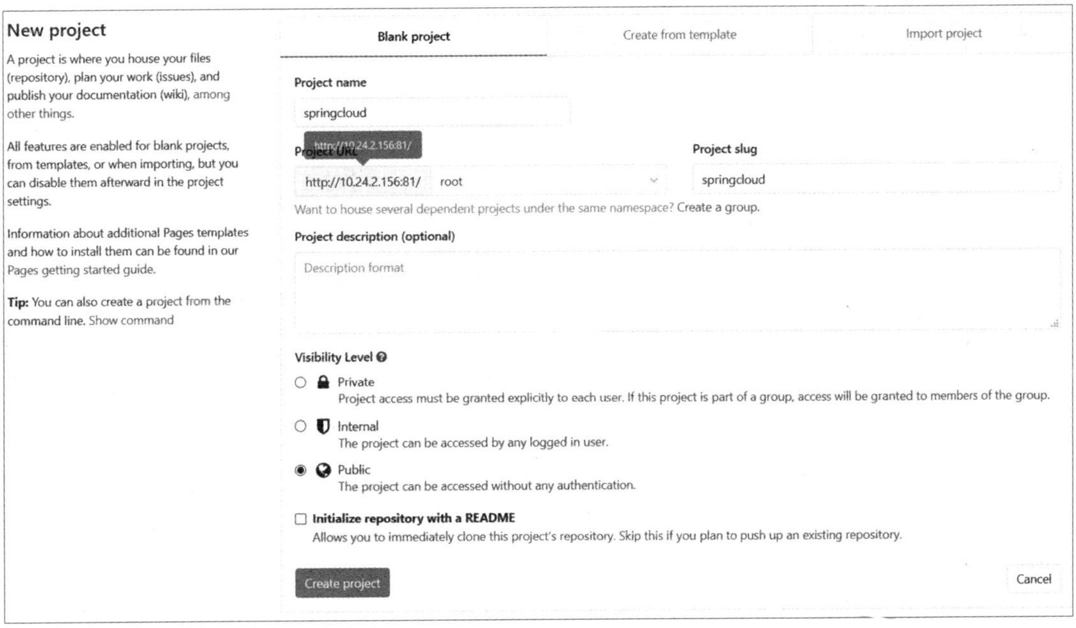

图 9-11　创建项目

单击"Creat project"按钮进入项目，如图 9-12 所示。

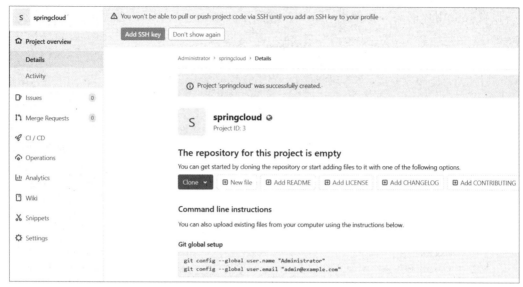

图 9-12　项目界面

回到 master 节点，push 源代码到 GitLab 的 springcloud 项目，命令如下：

[root@master ~]# yum install -y git
[root@master ~]# cd /opt/springcloud/
[root@master springcloud]# git config --global user.name "administrator"
[root@master springcloud]# git config --global user.email "admin@example.com"
[root@master springcloud]# git remote remove origin
[root@master springcloud]# git remote add origin http://192.168.200.20:81/root/springcloud.git
[root@master springcloud]# git add .
[root@master springcloud]# git commit -m "initial commit"
[root@master springcloud]# git push -u origin master
Username for 'http://192.168.200.20:81': root
Password for 'http://root@192.168.200.20:81':
Counting objects: 3192, done.
Delta compression using up to 4 threads.
Compressing objects: 100% (1428/1428), done.
Writing objects: 100% (3192/3192), 1.40 MiB | 0 bytes/s, done.
Total 3192 (delta 1233), reused 3010 (delta 1207)
remote: Resolving deltas: 100% (1233/1233), done.
To http://192.168.200.20:81/root/springcloud.git
 * [new branch]   master -> master
Branch master set up to track remote branch master from origin.

刷新网页，springcloud 项目中的文件已经更新了，如图 9-13 所示。

（4）配置 Jenkins 连接 GitLab

① 设置 Outbound requests。登录 GitLab 首页，如图 9-14 所示。

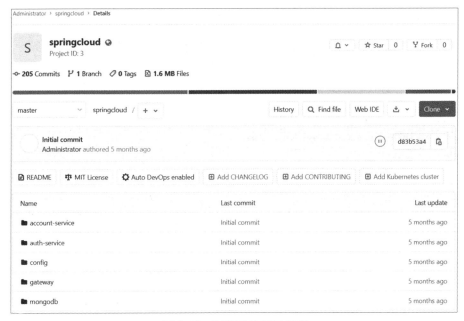

图 9-13　更新后的项目界面

图 9-14　GitLab 首页

单击管理区域的扳手图标,如图 9-15 所示。

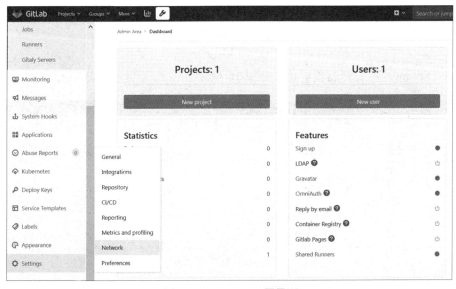

图 9-15　GitLab 配置界面

单击左侧导航栏的"Settings"→"Network",设置"Outbound requests",勾选"Allow requests to the local network from web hooks and services",如图 9-16 所示。

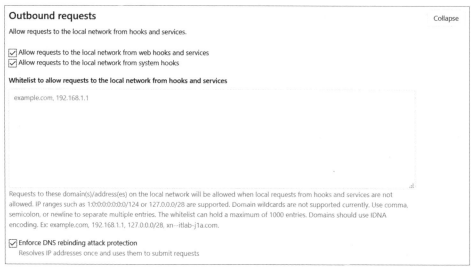

图 9-16　GitLab 配置界面

配置完成后保存。

② 创建 Gitlab API Token。单击 GitLab 用户头像图标,如图 9-17 所示。

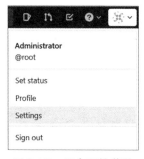

图 9-17　用户下拉菜单

选择"Settings",打开配置界面如图 9-18 所示。

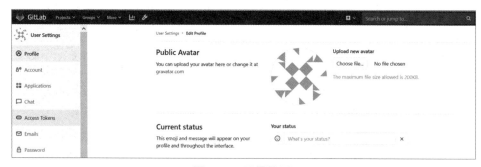

图 9-18　配置界面

单击左侧导航栏的"Access Tokens"按钮添加 Token,如图 9-19 所示。

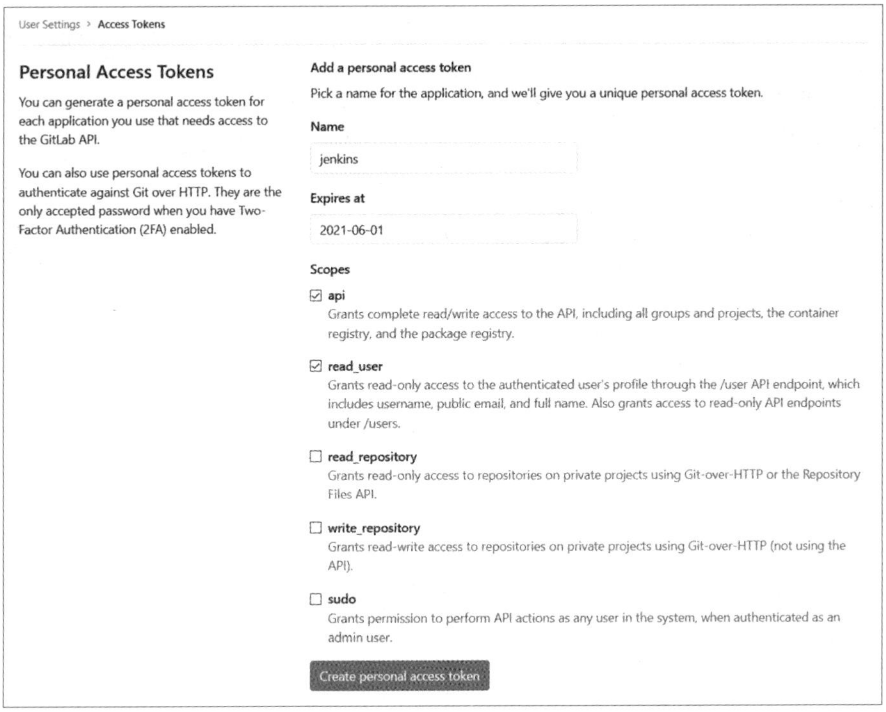

图 9-19　添加 Token 界面

单击"Create personal access token"按钮生成 Token，如图 9-20 所示。

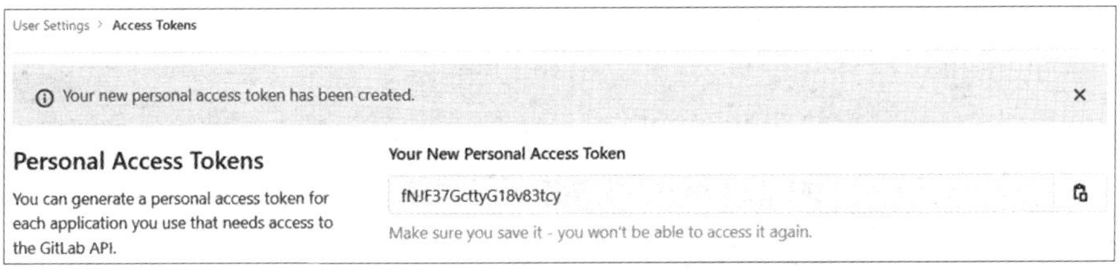

图 9-20　生成 Token 界面

复制 Token（fNJF37GcttyG18v83tcy），后面配置 Jenkins 时会用到。

③ 设置 Jenkins。登录 Jenkins 首页，单击"系统管理"→"系统配置"，配置 GitLab 信息，取消勾选"Enable authentication for '/project' end-point"，如图 9-21 所示。

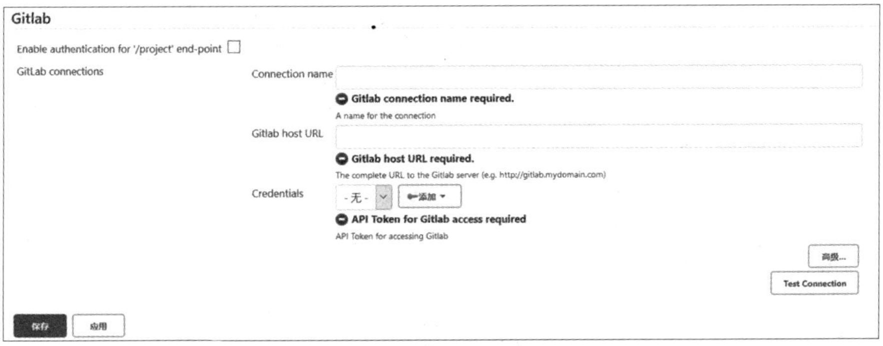

图 9-21　配置 GitLab 信息界面

167

单击"添加"→"Jenkins"添加认证信息，将 GitLab API token 填入，如图 9-22 所示。

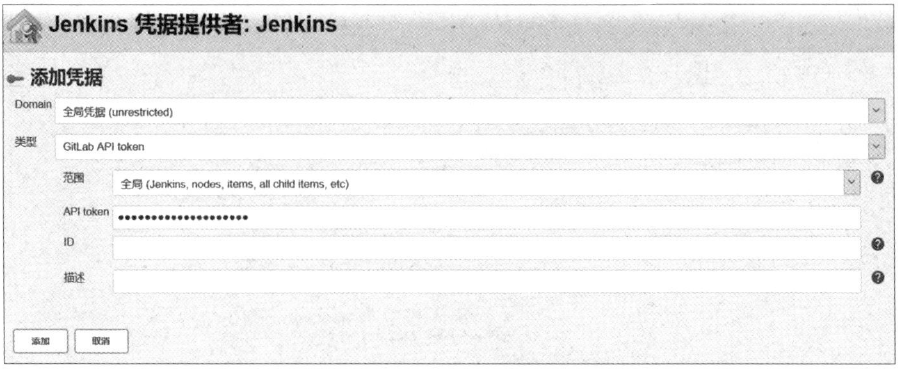

图 9-22　添加凭据界面

填入信息后，首先单击下方的"添加"按钮，然后单击右下角"Test Connection"按钮，如图 9-23 所示。

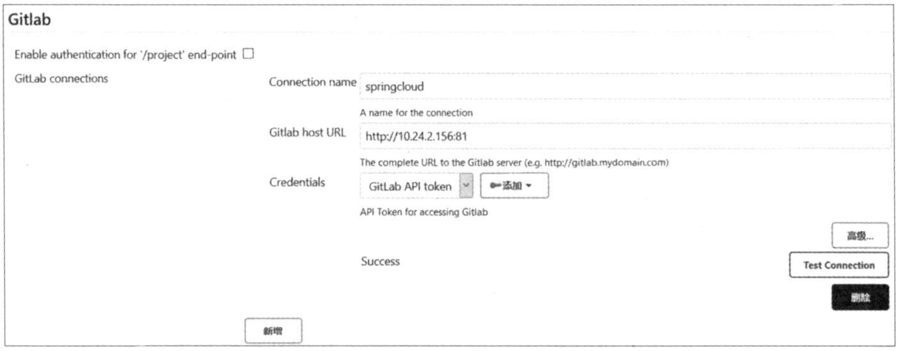

图 9-23　测试连接界面

至此配置 Jenkins 连接 GitLab 成功。

（5）配置 Jenkins 连接 maven

① 安装 maven。由于 Jenkins 是采用 Docker in Docker 的方式启动的，所以需要在 Jenkins 容器内安装 maven，命令如下：

```
[root@master ~]# cp -rf /opt/apache-maven-3.6.3-bin.tar.gz /home/jenkins_home/
[root@master ~]# docker exec -it jenkins bash
[root@344d4fa5b8ea:/]# tar -zxvf /var/jenkins_home/apache-maven-3.6.3-bin.tar.gz -C .
[root@344d4fa5b8ea:/]# mv apache-maven-3.6.3/ /usr/local/maven
[root@344d4fa5b8ea:/]# vi /etc/profile
export M2_HOME=/usr/local/maven    # 行末添加两行
export PATH=$PATH:$M2_HOME/bin
[root@344d4fa5b8ea /]# vi /root/.bashrc
# .bashrc
```

```
# User specific aliases and functions

alias rm='rm -i'
alias cp='cp -i'
alias mv='mv -i'

# Source global definitions
if [ -f /etc/bashrc ]; then
        . /etc/bashrc
source /etc/profile   # 添加本行
fi
```

退出容器重新进入，命令如下：

```
[root@344d4fa5b8ea /]# mvn -v
Apache Maven 3.6.3 (cecedd343002696d0abb50b32b541b8a6ba2883f)
Maven home: /usr/local/maven
Java version: 1.8.0_265, vendor: Oracle Corporation, runtime: /usr/lib/jvm/java-1.8.0-openjdk-1.8.0.265.b01-0.el8_2.x86_64/jre
Default locale: en_US, platform encoding: ANSI_X3.4-1968
OS name: "linux", version: "3.10.0-862.2.3.el7.x86_64", arch: "amd64", family: "unix"
```

② 连接 maven。登录 Jenkins 首页，单击左侧导航栏中的"系统管理"→"全局工具配置"，打开全局工具配置界面，如图 9-24 所示。

图 9-24　全局工具配置界面

单击"新增 Maven"按钮，打开新增 Maven 界面，如图 9-25 所示。

图 9-25 新增 Maven 界面

取消勾选"自动安装"选项，填入 Maven 名称和安装路径，配置完成后单击左下角的"应用"按钮即可。至此，关于 CI/CD（持续集成与持续发布）的基础环境准备完毕，接下来要编写流水线脚本，进行配置与触发。

2. CI/CD 环境配置与触发

（1）配置 CI/CD

① 新建任务。登录 Jenkins 首页，单击左侧导航栏中的"新建任务"按钮，打开新建任务界面，如图 9-26 所示。

图 9-26 新建任务界面

选择构建一个"流水线"，单击"确定"按钮，配置构建触发器，如图 9-27 所示。

记录下 GitLab webhook URL 的地址（http://192.168.200.20:8080/project/springcloud），后期配置 webhook 时需要使用。配置流水线，如图 9-28 所示。

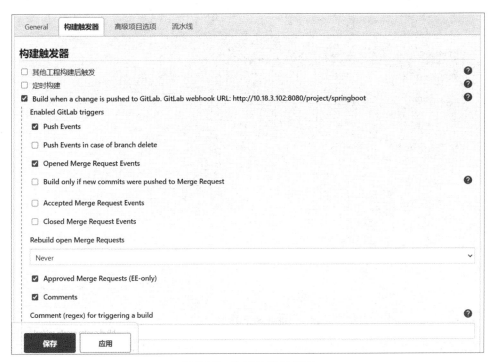

图 9-27 构建触发器界面

图 9-28 配置流水线界面

单击"流水线语法"按钮，打开如图 9-29 所示界面。

"示例步骤"选择"git：Git"，将 springcloud 项目地址填入"仓库 URL"。然后单击"添加"→"jenkins"选项添加凭据，如图 9-30 所示。"类型"选择"Username with password"，"用户名"和"密码"为 Gitlab 仓库的用户名和密码。

添加凭据后选择凭据，如图 9-31 所示。

图 9-29　配置流水线界面

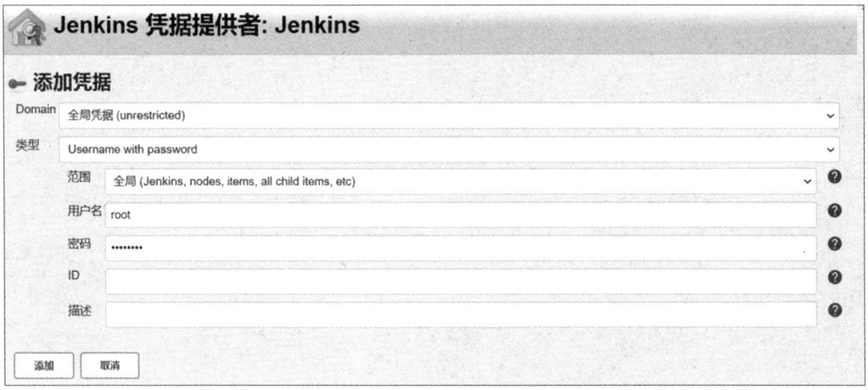

图 9-30　添加凭据界面（1）

图 9-31　添加凭据界面（2）

单击"生成流水线脚本"按钮，如图 9-32 所示。

图 9-32　生成流水线脚本

记录生成的值，并将其写入流水线脚本中，完整的流水线脚本如下：

```
node{

    stage('git clone'){
        //check CODE
        git credentialsId: 'f30e738e-7795-4348-b3ee-eb73cd274add', url: 'http://192.168.200.20:81/root/springcloud.git'
    }
    stage('maven build'){
        sh '''/usr/local/maven/bin/mvn package -DskipTests -f /var/jenkins_home/workspace/springcloud'''
    }
    stage('image build'){
        sh '''
            echo $BUILD_ID
            docker build -t 192.168.200.20/springcloud/gateway:$BUILD_ID -f /var/jenkins_home/workspace/springcloud/gateway/Dockerfile  /var/jenkins_home/workspace/springcloud/gateway
            docker build -t 192.168.200.20/springcloud/config:$BUILD_ID -f /var/jenkins_home/workspace/springcloud/config/Dockerfile  /var/jenkins_home/workspace/springcloud/config'''
    }
    stage('test'){
        sh '''docker run -itd --name gateway 192.168.200.20/springcloud/gateway:$BUILD_ID
        docker ps -a|grep springcloud|grep Up
        if [ $? -eq 0 ];then
            echo "Success!"
            docker rm -f gateway
        else
            docker rm -f gateway
            exit 1
        fi

'''
```

```
        }
        stage('upload registry'){
            sh '''docker login 192.168.200.20 -u=admin -p=Harbor12345
                docker push 192.168.200.20/springcloud/gateway:$BUILD_ID
                docker push 192.168.200.20/springcloud/config:$BUILD_ID'''
        }
        stage('deploy Rancher'){
            //执行部署脚本
            sh 'sed -i "s/sqshq\\/piggymetrics-gateway/192.168.200.20\\/springcloud\\/gateway:$BUILD_ID/g" /var/jenkins_home/workspace/springcloud/yaml/deployment/gateway-deployment.yaml'
            sh 'sed -i "s/sqshq\\/piggymetrics-config/192.168.200.20\\/springcloud\\/config:$BUILD_ID/g" /var/jenkins_home/workspace/springcloud/yaml/deployment/config-deployment.yaml'
            sh 'kubectl create ns springcloud'
            sh 'kubectl apply -f /var/jenkins_home/workspace/springcloud/yaml/deployment/gateway-deployment.yaml --kubeconfig=/root/.kube/config'
            sh 'kubectl apply -f /var/jenkins_home/workspace/springcloud/yaml/deployment/config-deployment.yaml --kubeconfig=/root/.kube/config'
            sh 'kubectl apply -f /var/jenkins_home/workspace/springcloud/yaml/svc/gateway-svc.yaml --kubeconfig=/root/.kube/config'
            sh 'kubectl apply -f /var/jenkins_home/workspace/springcloud/yaml/svc/config-svc.yaml --kubeconfig=/root/.kube/config'
        }
}
```

脚本中所有 IP 地址均为 Harbor 仓库的地址，然后在网页中写入完整的流水线脚本，如图 9-33 所示，完成后单击"应用"按钮。

图 9-33　写入流水线脚本

② 开启 Jenkins 匿名访问。登录 Jenkins 首页，单击"系统管理"→"全局安全配置"，配置授权策略为允许匿名用户访问，即选中"匿名用户具有可读权限"，如图 9-34 所示。

图 9-34　授权策略界面

③ 配置 Webhook。登录 GitLab，进入 springcloud 项目，单击左侧导航栏"Settings"→"Webhooks"，将前面记录的 GitLab webhook URL 地址填入"URL"处，禁用 SSL 认证，如图 9-35 所示。

图 9-35　禁用 SSL 认证界面

单击"Add webhook"按钮添加 Webhook，完成后的界面如图 9-36 所示。

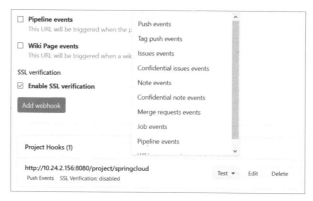

图 9-36　添加 Webhook 界面

单击右侧"Test"→"Push events"进行测试，如图 9-37 所示。

图 9-37　测试界面

结果返回 HTTP 200 则表明 Webhook 配置成功。

④ 创建仓库项目。登录 Harbor，新建项目 springcloud，访问级别设置为公开，创建完成后的页面如图 9-38 所示。

图 9-38　Harbor 仓库首页

进入项目查看镜像列表，如图 9-39 所示，此时为空，无任何镜像。

图 9-39　Harbor 仓库列表

（2）触发 CI/CD

① 触发构建。在 master 节点上传代码触发自动构建，具体命令如下：

```
[root@master ~]# docker cp /opt/repository/ jenkins:/root/.m2/
[root@master ~]# docker restart jenkins
jenkins
[root@master ~]# cd /opt/springcloud/
[root@master springcloud]# git add .
[root@master springcloud]# git commit -m "Initial commit"
# On branch master
nothing to commit, working directory clean
[root@master springcloud]# git push -u origin master
Username for 'http://10.24.2.156:81': root
Password for 'http://root@10.24.2.156:81':
Branch master set up to track remote branch master from origin.
Everything up-to-date
```

② Jenkins 查看。登录 Jenkins 界面，可以看到 springcloud 项目已经开始构建，如图 9-40 所示。

图 9-40　Jenkins 首页

单击项目名称查看流水线阶段视图，如图 9-41 所示。

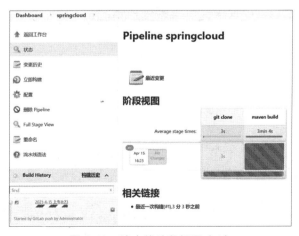

图 9-41　流水线阶段视图（1）

单击右侧"#1"按钮可查看控制台输出，此处会显示构建的详细进程，如图 9-42 所示。

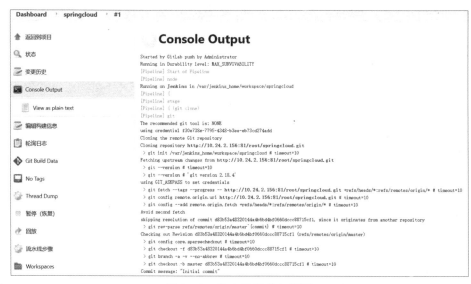

图 9-42　构建详细信息界面

构建完成后的控制台输出如图 9-43 所示。

图 9-43　构建完成后的控制台输出

返回项目查看流水线阶段视图，如图 9-44 所示。

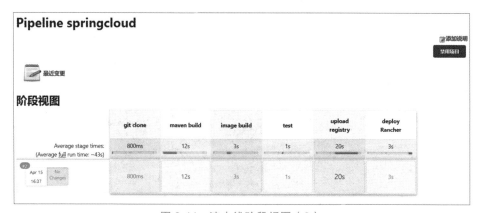

图 9-44　流水线阶段视图（2）

③ Harbor 查看。进入 Harbor 仓库 springcloud 项目查看镜像列表，可以看到已自动上传了一个 gateway 镜像，如图 9-45 所示。

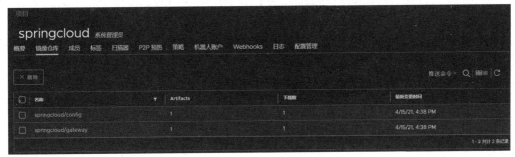

图 9-45　Harbor 镜像列表

④ Kubernetes 查看。Pod 的启动较慢，需等待 3～5 分钟。查看 Pod，命令如下：

```
[root@master ~]# kubectl -n springcloud get pods
NAME                          READY   STATUS    RESTARTS   AGE
config-6c988c4dc5-2522c       1/1     Running   0          21m
gateway-6545fc58c5-d6rgn      1/1     Running   0          21m
```

查看 service，命令如下：

```
[root@master ~]# kubectl -n springcloud get service
NAME      TYPE       CLUSTER-IP      EXTERNAL-IP   PORT(S)          AGE
config    NodePort   10.101.42.47    <none>        8888:30015/TCP   22m
gateway   NodePort   10.100.62.39    <none>        4000:30010/TCP   22m
```

通过端口 30010 访问服务，如图 9-46 所示。

图 9-46　应用首页

至此，完整的 CI/CD 流程就完成了。

归纳总结

通过本单元内容的学习，对 CI/CD 的使用场景、工作流程、解决工作中的痛点有了一定的认识，也掌握了 CI/CD 服务的优点与优势。在今后的工作中，如果使用了 CI/CD，会大大减少开发人员的工作量，也会使整个开发流程更加规范，提高了效率。CI/CD 是 DevOps 工具中不可或缺的理念与方法，有了 CI/CD 和自动化，可以进行频繁的集成、发布，到时候手动操作造成的错误等问题就被消除了。关于更多 CI/CD 的最佳实践，读者可以自行查找资料进行更深入的学习。

课后练习

一、判断题

1. Kubernetes 查看当前支持的 API 版本的命令是 kubectl get pods。（   ）
2. PV 是 Kubernetes 用来进行存储资源声明的。（   ）

二、单项选择题

1. 下面（   ）不属于容器云技术的范畴。

A. Docker　　　　B. Kubemetes　　　　C. OpenShift　　　　D. OpenStack

2. Docker 容器技术有（   ）种网络模式。

A. 1　　　　B. 2　　　　C. 3　　　　D. 4

三、多项选择题

下面关于 Kubernetes 中 Pod 概念的说法中，不正确的是（   ）。

A. Kubernetes 为每个 Pod 都分配了唯一的 IP 地址
B. Kubernetes 要求底层网络支持集群内任意两个 Pod 之间的 TCP/IP 直接通信
C. 静态的 Pod 存放在 Kubernetes 的 etcd 中
D. 当 Pod 里的某个容器停止时，Kubernetes 不会重启这个 Pod

开源容器云平台 OpenShift

技能训练

1. 在 Kubernetes 节点启动 Jenkins 容器，并配置 Maven 环境。
2. 构建 CI/CD 流水线项目，并配置触发器，完成流水线镜像推送至 Harbor 项目。

开源企业容器云的平台部署　　开源企业容器云的业务系统部署　　开源企业容器云的业务系统运维

单元 9　基于 K8S 的 CI/CD 服务

原生 Kubernetes 容器云平台运维

开源企业容器云的平台部署

原生 Kubernetes 容器云平台基本测试使用

开源企业容器云的业务系统部署

开源企业容器云的业务系统运维

原生 Kubernetes 容器云平台运维

反侵权盗版声明

　　电子工业出版社依法对本作品享有专有出版权。任何未经权利人书面许可，复制、销售或通过信息网络传播本作品的行为，歪曲、篡改、剽窃本作品的行为，均违反《中华人民共和国著作权法》，其行为人应承担相应的民事责任和行政责任，构成犯罪的，将被依法追究刑事责任。

　　为了维护市场秩序，保护权利人的合法权益，我社将依法查处和打击侵权盗版的单位和个人。欢迎社会各界人士积极举报侵权盗版行为，本社将奖励举报有功人员，并保证举报人的信息不被泄露。

举报电话：（010）88254396；（010）88258888
传　　真：（010）88254397
E-mail：　dbqq@phei.com.cn
通信地址：北京市海淀区万寿路 173 信箱
　　　　　电子工业出版社总编办公室
邮　　编：100036